SALES IS DEAD

The Rise of AI

Chris J. Martinez

J. JOSEPH GROUP

Published in the United States by J. Joseph Group, LLC
Round Rock, Texas

Sales Is Dead / Chris J. Martinez. — 1st ed.

Library of Congress Control Number:
Paperback ISBN: 978-1-7355869-6-0
eBook ISBN: 978-1-7355869-7-7

Cover design by JetLaunch.net

www.ChrisJosephMartinez.com

ALSO BY CHRIS J. MARTINEZ

Driving Sales: What It Takes to Sell 1000+ Cars Per Month

The Drive to 30: Your Ultimate Guide to Selling More Cars than Ever

The Unfair Advantage: Digital Marketing Principles that Will Explode the Growth of an Auto Dealership

The Rainmaker: Fundamentals of the Car Dealer's Desk Manager

The Closer: The Automotive Professional's Guide to Closing the Deal

The Digital Marketing Blueprint: A Car Dealer's Guide to Selling More Cars

When I Grow Up, I Want To Be a Car Salesperson! (children's book)

To my wife, Veronica—thank you for your unwavering belief in me and for always supporting me to perform at my best. Your strength and partnership in raising our four children mean the world to me. I love you!

And to Jazmin, Janelle, Julianna, and Christian—know that anything is possible. I love you all to the moon and back.

CONTENTS

INTRODUCTION..9

PART 1

The History and Evolution of Sales 13

    Ch 1: Bartering to Closing Deals 15

    Ch 2: The Golden Age of Sales Professionals 25

    Ch 3: The Art of Selling.................................... 33

    Ch 4:Negotiation Mastery 43

PART 2

Why the Sales Professional is Still Relevant.......... 51

    Ch 5: Sales Expertise in an AI World 53

    Ch 6: From Amateur to Master 61

    Ch 7: The Human Element................................. 71

    Ch 8: Thriving in AI-Driven Sales Landscape..... 79

PART 3:

AI's Takeover and Transformation of Sales.......... 83

    Ch 9: AI in Sales 85

    Ch 10: The Role of Data in AI-Driven Sales 93

    Ch 11: From Zero to Guru................................... 99

    Ch 12: The Right Tools and Platforms.............. 107

    Ch 13: The Future of Sales with AI 113

PART 4

The New Sales Landscape................................. 121

Ch 14: Platforms Changing the Game 123

Ch 15: Inbound and Outbound Reinvented 133

Ch 16: Automation Meets Personalization 141

Ch 17: Cost Efficiency...147

PART 5

Capitalizing on AI Sales Platforms......................155

Ch 18: Mastering AI Platforms............................157

Ch 19: Training with IgniteUps.AI......................163

Ch 20: The Skills to Sell More with AI..............171

Ch 21: From Implementation to Profit 179

PART 6

The Future of Sales in an AI-Driven World.........185

Ch 22: The Next Decade of Sales........................187

Ch 23: Navigating the Shift 195

Ch 24: Human Creativity Meets AI Precision....197

Ch 25: Building a Winning Sales Strategy 205

ACKNOWLEDGEMENTS......................................213

APPENDICES 217

The End of an Era
Why Traditional Sales is Dying

For decades, sales has been at the core of every successful business, and the iconic salesperson—charismatic, persuasive, and armed with a Rolodex of clients—was once the backbone of industries across the globe. Entire organizations thrived on the abilities of skilled professionals who could close deals, nurture relationships, and generate profits. But the world has changed and so has the game.

In today's fast-paced, tech-driven world, traditional sales methods are quickly becoming obsolete. The shift isn't just about digital tools or social media; it's about something far more revolutionary. Artificial intelligence (AI) is taking over at a rapid pace, and it's reshaping the sales landscape in ways we've only just begun to understand.

The reality is stark: sales as we know it is dead.

But that doesn't mean that the *art of selling* is irrelevant. In fact, it's more important than ever. The question isn't whether sales professionals are still needed—it's about how they can evolve to thrive in an AI-driven world. The rise of AI doesn't mark the end of salespeople; it marks the end of those unwilling to adapt.

THE OPPORTUNITY OF
A LIFETIME
AI'S ROLE IN SHAPING THE FUTURE

AI isn't just a new tool; it's a game changer. It has the power to transform anyone—yes, even the most inexperienced amateur—into a sales guru. With platforms like IgniteUps.AI, Spearphish.io, and SalesAI, companies can create inbound and outbound AI agents that handle calls, emails, and texts with unmatched precision and efficiency. Imagine being able to replace an entire call center with AI agents that costs 1% of what a single human employee does, without sacrificing quality.

This shift doesn't just save money; it changes the way we think about sales entirely. AI platforms can automate the repetitive, data-heavy tasks that bog down human sales teams, leaving more room for creativity, strategy, and personalization. They give businesses the ability to scale their operations, reach new markets, and close deals faster than ever before.

But here's the catch: To succeed in this new landscape, you need to understand how AI works, how to use it, and—most importantly—how to sell with it. Without the foundational skills of salesmanship and negotiation, even the most advanced AI platform will fall short. The human elements—vision, strategy, and creativity—will always be critical.

> "It's not about displacing humans; it's about humanizing the digital experience."
>
> —*Rob Garf, Vice President and General Manager at Salesforce Retail*

WHY THIS BOOK MATTERS

This book serves as not just a wake-up call but a guide for those who want to lead, not follow, in the new era of sales. It's about learning from the past, embracing the future, and leveraging the tools that will define success in the decades to come.

In the pages ahead, we'll explore the history of sales and its evolution into an art form. We'll discuss why the skills of persuasion, negotiation, and relationship-building still matter. Then, we'll dive deep into how AI is transforming the sales landscape, from today's simple platforms to the fully autonomous sales operations of the future.

Whether you're a seasoned sales professional or an entrepreneur looking to scale your business, this book will show you how to harness the power of AI to create a competitive advantage. Together, we'll explore how to blend the timeless art of selling with cutting-edge AI technology to ensure you don't merely survive but you also thrive in a world where sales is dead—and then reborn.

Welcome to the future. Let's get started.

PART 1

The History and Evolution of Sales

CHAPTER 1

Bartering to Closing Deals
The Origins of Selling

Before skyscrapers, boardrooms, and global economies, there was basic trade. At its core, sales is about solving problems and meeting needs, and this fundamental truth has been with humanity since the dawn of civilization. Long before there were slick sales pitches or closing techniques, there were simple exchanges—bartering goods for survival.

Sales, in its most primitive form, began as an act of survival. If you needed something you couldn't create or harvest yourself, you turned to others to meet that need. What seems like a basic concept today was, in fact, the start of an evolutionary process that would shape economies, societies, and industries. In this chapter, we'll explore the earliest origins of selling, the evolution of trade, and how the principles established thousands of years ago still underpin modern sales strategies.

THE DAWN OF BARTERING HUMANITY'S FIRST SALES PITCH

Bartering was humanity's first economic system, a way to exchange goods and services without the need for money. It's easy to picture early humans trading a flint knife for a bundle of firewood or a few animal skins for a supply of dried meat. These exchanges weren't just about survival—they were the first instances of what we now call sales.

Bartering required more than just goods to trade; it demanded solid communication and persuasion skills. If one person had surplus grain and needed meat, they had to convince a hunter that the grain they had for "sale" was worth giving up some of their hard-earned catch. The process was inherently relational. In short, trust was built, value was communicated, and agreements were reached.

One of the key challenges of bartering was determining equivalency. How many fish were worth a bushel of wheat? What if one party didn't feel they were getting a fair deal? This introduced the need for negotiation, a skill that remains central to sales today.

There were two primary lessons learned from bartering: understanding needs and building trust. The best barterers understood not just what they needed but what the other party valued. Successful exchanges required mutual respect and trust, which continue to be foundational elements of any sales relationship.

While the barter system worked for small communities, it became cumbersome as societies expanded. The limitations of bartering—such as finding someone who wanted exactly what you had to offer—led to the next stage of sales evolution.

ANCIENT CIVILIZATIONS AND THE RISE OF TRADE

As human societies grew more complex, so did the systems of trade. Ancient civilizations like Mesopotamia, Egypt, and China were some of the first to develop organized marketplaces. These bustling hubs became the epicenters of commerce, where farmers, artisans, and traders gathered to buy and sell goods.

In Mesopotamia, the practice of trade was heavily documented. If you can imagine it, they used clay tablets as receipts for transactions, whereas today we get digital receipts in an instant on our mobile devices. Nevertheless, the innovation made commerce more efficient and, at the same time, introduced the concept of accountability. Traders could no longer rely solely on verbal agreements; they needed to prove they'd met their promises and obligations with written records.

Egyptian marketplaces, often located near temples, were highly organized. Merchants sold everything from textiles to spices, and their ability to persuade buyers became a critical skill. Selling was no longer just about the product—it was about creating an experience. Egyptian sellers used demonstrations, storytelling, and even humor to attract customers.

The Greeks and Romans took salesmanship to another level. In ancient Greece, the agora (marketplace) was where philosophers debated, and merchants sold their goods. Sellers had to be skilled in rhetoric to compete in such a lively environment. Meanwhile, Roman traders developed sophisticated negotiation tactics and marketing techniques. They used signs, public demonstrations, and even discounts to entice buyers.

Several notable innovations developed during the times of ancient trade. For one, Roman merchants were among the first to incorporate advertising via the use of slogans and signage to promote their goods. They also learned to identify buyer needs and adapt their pitch accordingly. All of those lessons still hold value today, as the ability to communicate effectively, build relationships, and adapt to the customer's needs remain at the heart of successful selling.

THE BIRTH OF CURRENCY
REVOLUTIONIZING SALES

The birth of currency laid the foundation for modern commerce, introducing many of the principles that govern sales today.

Around 600 BC, the ancient Lydians introduced the first form of currency: coins made of precious metals. This marked a turning point in the history of sales. With the invention of money, trade became more efficient, as people no longer had to find someone who wanted their specific goods. They could simply use coins to purchase what they needed.

Currency transformed the concept of value. For the first time, goods and services had a standardized price, which made transactions simpler and more predictable. However, this also introduced new complexities. Sellers now had to determine pricing strategies, evaluate costs, and compete with others offering similar products.

The introduction of currency also gave rise to early forms of credit. In ancient Babylon, for instance, merchants would extend loans to buyers, allowing them to pay for goods over time. This innovation not only expanded the reach of sales but also introduced the concept of financial risk.

The introduction of a standardized currency changed sales by allowing for standardized pricing (which enabled sellers to clearly communicate the value of their goods), expanded markets (allowing trade to flourish over larger distances), and increased competition (with standardized pricing, sellers had to find other ways to differentiate themselves).

THE INTRODUCTION OF TRADE ROUTES

With the rise of organized economies came the development of trade routes, connecting distant regions with one another and facilitating the exchange of goods, cultures, and ideas.

The Silk Road, for example, linked China with Europe, enabling the trade of not only silk but also spices and precious metals. These routes both expanded the reach of sales and elevated the role of the salesperson.

Merchants traveling along trade routes had to be more than just sellers—they also had to be diplomats, negotiators, and problem-solvers. They dealt with language barriers, cultural differences, and the constant threat of theft or fraud. Successful merchants, therefore, were the ones who were able to adapt to new environments and build trust with diverse groups of people.

KEY LESSONS FROM
ANCIENT TRADE ROUTES

Adaptability

Merchants had to adjust their strategies based on the culture and preferences of their customers.

Resilience

Traveling and trading along these routes required persistence and ingenuity.

Value of Relationships

Long-term success depended on one's ability to build strong relationships with buyers and other traders.

By the time the Industrial Revolution arrived (1760-1840), the seeds of modern salesmanship had already been sown. The principles of trust, negotiation, and communication that emerged in ancient times became the foundation for the sales techniques we recognize today. However, the lessons of history remind us that sales isn't just about transactions. It's also about relationships, adaptability, and creating value.

As we move forward in this book, these historical foundations will serve as a foundational guide.

Because whether you're using AI-driven platforms or negotiating face-to-face, an understanding of the timeless principles of selling remains critical.

The story of sales is one of constant evolution, driven by innovation and the ever-changing needs of humanity. From bartering in small villages to trading across continents, the sales process has always been about connecting people and solving problems.

Today, we stand on the brink of yet another transformation: the rise of artificial intelligence. But as revolutionary as AI may be, it doesn't replace the core principles that have defined sales for thousands of years. Instead, it enhances them, giving us new tools to build relationships, communicate value, and close deals.

Before we go much further, it's important to explore the rise of the sales professional and why their role remains vital in an AI-driven world. Contrary to the common fear that salespeople will be fully replaced by the evolution of AI technology, the human touch will remain irreplaceable, and it's critical to understand how that affects you.

"AI is pushing us to do what we should be doing anyway: the creation of more humanistic service jobs."

—*Dr. Kai-Fu Lee, Chairman and CEO of Sinovation Ventures*

CHAPTER 2

The Golden Age of Sales Professionals

The twentieth century marked the transition from survival-based bartering to professional selling as we know it today. For one thing, with the rise of industrialization, mass production, and consumer culture, the role of the salesperson became more defined and essential. Sales professionals became the bridge between manufacturers and consumers, armed with skills, strategies, and tools to close deals and generate revenue.

From door-to-door sales to the practices of corporate giants like IBM and Xerox, the way the rise of structured sales training turned sales into a respected and coveted career is both fascinating and enlightening.

INDUSTRIALIZATION AND THE
BIRTH OF MODERN SALES

The Industrial Revolution was a pivotal period in the history of sales. Factories were able to churn out products faster than ever before, creating an abundance of goods that needed to be sold. Further, businesses could no longer rely solely on local markets and, as a result, needed sales professionals who could expand their reach and build the trust necessary to persuade new customers to buy.

One of the earliest examples of professional selling emerged in the form of the traveling salesman. The 1920s to 1960s is often considered the golden age in the area of door-to-door sales. Companies like Avon, Fuller Brush, and Electrolux sent armies of salespeople directly to consumers' homes. These individuals became household names, representing not just products but trust and convenience.

Known for their charisma, persistence, and ability to connect with people from all walks of life, these individuals crisscrossed towns and entire cities, selling products like sewing machines, encyclopedias, and cookware, and, in the process, laying the groundwork for modern sales professionals by demonstrating key principles that have stood the test of time.

Relationships were at the forefront of their businesses, as their success ultimately depended on their ability to form personal connections with customers. They also had to take the time to educate customers on why their product was worth the investment. And of course, rejection was as common then as it is now, so it was essential that they be able to persist in the face of objections and outright refusals.

Door-to-door sales thrived *because* it was personal, which is an incredibly important aspect to note. Customers could see the product in action via a tailored presentation, ask questions, and receive customized responses. Salespeople became more and more skilled at identifying customer needs and demonstrating how their products met those needs. For example, a Fuller Brush salesperson didn't just sell cleaning tools—they sold cleanliness, efficiency, and ease of use. Similarly, Avon representatives offered not just cosmetics but the promise of beauty and confidence.

Yet, even after lengthy in-person demonstrations, door-to-door salespeople received rejections. Therefore, every interaction had to be customized based on the customer's reactions and needs. And, because customers were inviting salespeople into their homes, trust was paramount. Think about it: today, doorbells aren't often even rung by door-to-door salespeople, as homeowners have affixed No Soliciting signs to their door or, upon seeing someone they don't recognize on their Ring camera, opted to simply not answer.

The mid-twentieth century saw the rise of corporate giants that transformed sales into a structured profession. Companies like IBM and Xerox developed formal sales methodologies, turning selling into a science.

IBM and the Consultative Sales Model

IBM was a trailblazer in developing a customer-focused sales approach. Rather than pushing products, IBM's salespeople were trained to act as consultants, identifying customer needs and offering tailored solutions. This approach revolutionized the way businesses viewed sales, shifting the focus from closing deals to building long-term relationships.

Xerox and the Birth of Sales Training

Xerox is often credited with creating the first formal sales training program. The Xerox Professional Selling Skills program taught salespeople how to ask the right questions, listen actively, and position their product as the solution to the customer's problem. This program became the gold standard for sales training, influencing countless other companies, including SAP and Microsoft.

SAP (an enterprise software company) implemented a structured sales methodology inspired by Xerox's PSS framework, focusing on understanding customer pain points and offering tailored solutions, and Microsoft's sales teams have historically used training methods that align with Xerox's principles, emphasizing solution-selling and customer-oriented approaches to drive software and services adoption.

*****

During this era, the salesperson became a symbol of ambition and success. Movies, television, and literature portrayed salespeople as charming, persuasive individuals who could talk their way into anything. While some depictions were unflattering (e.g., Willy Loman in *Death of a Salesman*), many celebrated the salesperson's grit and determination.

Notable individuals such as Zig Ziglar and Mary Kay Ash turned sales into an aspirational career, proving that with the right mindset and skills, anyone could succeed.

Ziglar, one of the most famous sales trainers, inspired generations of sales professionals with his motivational speeches and practical advice. He taught that sales was about serving others *and* solving problems. And Ash, founder of Mary Kay Cosmetics, built a multi-billion-dollar cosmetics business by empowering women to become sales leaders. Her focus on recognition and personal development revolutionized the industry.

Secrets of success sales also began making their way into books. Dale Carnegie's 1936 book, *How to Win Friends and Influence People*, became a bible for sales professionals. Within the pages of this classic self-improvement book that's since sold millions of copies, Carnegie emphasized the importance of qualities such as empathy, active listening, and understanding human psychology. His teachings resonated with salespeople precisely because they focused on building genuine connections rather than relying on manipulative tactics.

TOOLS OF THE TRADE
EARLY INNOVATIONS IN SALES

The golden age of sales saw the introduction of tools and technologies that made selling more efficient. These innovations laid the groundwork for today's digital sales tools, from CRMs to AI-powered platforms.

For one, there was the invention of the telephone, which allowed salespeople to reach more customers without having to leave their office. Cold calling became a staple of sales strategies. And companies like Sears began using mail-order catalogs to showcase their products, giving salespeople a visual tool to enhance their pitches. Business cards also became an essential tool for building credibility and leaving a lasting impression.

LESSONS FROM THE GOLDEN AGE

Focus on the Customer

The best salespeople understood that the customer's needs came first.

Persistence Pays Off
Whether knocking on doors or making cold calls, resilience was key to success.

Continuous Learning
Sales training programs emphasized the importance of honing skills and staying ahead of the competition.

Perhaps the most enduring lesson: Sales is fundamentally about people. While tools, techniques, and philosophies may evolve, the ability to connect with others will always be at the heart of selling.

The golden age of sales professionals was a time of innovation, ambition, and transformation. It turned selling into a respected profession and laid the foundation for the modern sales industry. And the principles developed during this era—including relationship-building, persistence, and continuous learning—remain as relevant as ever.

While technology can enhance efficiency and scale, it's the human touch that will continue to drive trust and connection. So let's dive deeper into the art of selling, exploring the timeless skills that have built empires and transformed industries.

CHAPTER 3

The Art of Selling
Skills That Built Empires

Sales is not a transactional activity; it's an art. At its heart, selling is about human connection—understanding a person's needs, presenting a solution, and building trust to create a lasting relationship. While technology and innovation have transformed the ways in which we sell, the core skills that drive successful sales remain timeless. These are the skills that have built empires, turned startups into global giants, and defined industries.

The art of selling is, of course, not confined to the workplace—it's a life skill that enhances every area of your life. Whether you're negotiating a salary, persuading a friend to go try a new restaurant, or pitching an new idea to your team, the critical skills of that have built empires—listening, building value, exhibiting emotional intelligence, storytelling, and skillfully negotiating—are both timeless and universally applicable. They've helped businesses thrive in every era, and they will continue to do so in the age of AI.

HEARING VERSUS LISTENING

In sales, most people think the key to success is talking. But the truth is, great salespeople listen more than they speak. Listening allows you to uncover what your customer truly needs, which enables you to position your product or service as the perfect solution. And it's not just hearing—great salespeople must hone their ability to truly listen in order to understand what a customer is saying.

The best sales conversations start with understanding, not pitching.

Many sales professionals make the mistake of hearing instead of listening. Hearing is passive; it's waiting for your turn to speak. Listening, on the other hand, is active; it's about fully engaging with the customer and absorbing their words, tone, and emotions. There are a variety of techniques that encourage active listening.

First, there's the willingness to *ask open-ended questions.* Instead of asking questions that result in a simple "yes" or "no," ask questions that prompt the customer to share more. For example, "What's the biggest challenge you're currently facing?" or "How have you been handling this issue so far? What's working or not working for you?"

Second, be sure to *repeat key points* to show that you've understood what the customer is communicating, e.g., "So you're saying that reliability is your top priority?"

Third, *pause before responding*, giving the customer space to elaborate. Silence can encourage them to feel heard and to share more details.

When customers feel heard, they feel valued. Listening shows that you care about their needs, not just the sale.

Trust is the foundation of every successful transaction, and listening is how you earn it.

COMMUNICATING VALUE
GOING BEYOND FEATURES

A product's or service's value isn't about its features or specifications; it's about how it solves a problem or fulfills a desire. People don't buy products—they buy solutions, emotions, and outcomes. People don't buy luxury cars for the leather seats; they buy the prestige and confidence that come with driving a luxury vehicle. Similarly, people don't buy software for its interface; they buy the efficiency and time savings it provides.

There are three core ways to build value in your role as a salesperson. First, make sure you *understand the customer's pain points*. Ask questions that uncover their frustrations, challenges, and goals. Next, *connect to your product's features and benefits*. Don't just describe what your product does; explain how it improves their life or business based on their pain points. Finally, *customize your pitch*. Take the time to tailor your presentation to focus on the aspects of your product that matter most to your customer.

If you don't believe in your product, neither will your customer. Confidence is contagious, so speak with authority, share success stories, and demonstrate your expertise to make the customer feel they're making the right choice.

EMOTIONAL INTELLIGENCE
SELLING BEYOND LOGIC

Research shows that most buying decisions are emotional, even when they seem logical. People buy based on how they feel, then justify their decision with facts. Emotional intelligence—the ability to recognize, understand, and manage emotions—has always been and will continue to be a critical skill in sales. Thankfully, there are a number of proven ways to use emotional intelligence in sales.

Be sure to take the time to *find common ground* and create a connection with your customer. This could be as simple as commenting on shared interests or showing genuine curiosity about their situation. *Pay attention* to your customer's body language, tone of voice, and facial expressions. These often reveal more than words. And *don't neglect the importance of empathizing* with your customer. Put yourself in their shoes. Understand their concerns and reassure them that you're there to help.

CASE STUDY
Apple's Approach to Emotional Selling

Apple's success isn't just about innovative products; it's about how they make customers feel. Their advertising focuses on emotions—creativity, freedom, and self-expression—rather than just technical specs.

Apple's 1997 "Think Different" campaign tapped into customers' desire for creativity and individuality. The emotional appeal wasn't about tech specs but about belonging to a movement of visionaries. By associating its products with innovation, rebellion, and self-expression by featuring iconic figures such as Albert Einstein, Martin Luther King, Jr., and Amelia Earhart, Apple fostered deep emotional connections, making customers feel inspired and empowered.

THE ART OF STORYTELLING TURNING FEATURES INTO EXPERIENCES

Stories are, at once, memorable, emotional, and persuasive. They help customers visualize the impact of a product or service in their own lives. A great salesperson is a great storyteller, weaving in narratives that connect with the customer's desires and needs.

I know what you might be thinking, but telling a story is easier than you likely think. All you have to do is, start with your customer's main problem. This frames the story around a challenge your customer can easily relate to. Then, introduce the hero of the story (spoiler alert: the hero is your product or service). Explain how your product or service will solve their problem. Finally, paint the outcome for your customer. Highlight the positive results they will experience, being sure to focus on the transformation or success the customer will achieve. Remember, it's not about technical specs, it's about emotions. There's an emotion your customer is craving. Your goal is to help them see that your product or service will provide them with that emotion.

For example, a real estate agent might describe how a young family found their dream home. A software salesperson might share how a small business doubled its revenue with their platform. Or a car dealer might recount how a recent customer's new vehicle provided safety and peace of mind during a cross-country trip.

Stories engage both the logical and emotional parts of the brain. They make abstract concepts tangible and help customers see the value of your product in their own lives.

Timeless Storytelling Principles

1 — Discover Customer's Main Problem

2 — Introduce the Hero (Your Product/Service)

3 — Paint the Outcome

OVERCOMING OBJECTIONS AND CLOSING THE DEAL

Of course, you're bound to receive rejections along the way, and successful salespeople have learned how to overcome those objections. Here are some of the more common objections and suggestions on how to address them:

"It's too expensive."
Response: "I understand price is a concern. Let's talk about the value you're getting and how this investment can save you money in the long run."

"I need to think about it."
Response: "I completely understand. Can I help clarify anything we've discussed so you have all the information you need to make a decision?"

"I'm not sure it's the right fit."
Response: "That's fair. Let's revisit your goals and see if this is the best solution for your needs."

Once you've navigated your customer's objections, it's time to close the deal. While there are a number of techniques salespeople use to close deals, three have stood the test of time.

First, there's the *assumptive close*. This is when the salesperson acts as though the decision has already been made, saying something along the lines of "Should we schedule delivery for Friday or Monday?"

Next, there's the *urgency close*, which creates a sense of urgency for the buyer. You've no doubt been on the receiving end of statements such as "This offer is only available until the end of the week" and therefore understand the urgency they create.

To ensure an urgency close is ethical, make sure you tie it to real deadlines (e.g., limited stock, expiring bonuses) rather than false scarcity. (Customers are savvy about spotting the deal that's "always expiring in twenty-four hours" and it erodes trust. Be transparent about why acting now benefits the customer, and continue focusing on their unique needs, not pressure. Honesty builds trust, making urgency a helpful motivator rather than a manipulative tactic.

Last (but not least), there's the *takeaway close*. A takeaway close is a bit counterintuitive, as it involves suggesting that the customer walks away to encourage commitment by saying something like "I understand if this isn't the right time—let me know when you're ready."

The takeaway close builds trust by showing customers you respect their decision and won't pressure them if they aren't ready. It shifts the focus from selling to genuinely helping, making them feel in control. Even if they don't buy now, they'll remember your integrity and may return when they're ready because they value your honesty and professionalism.

While tools and platforms may change, the human element of sales will always be at the forefront. Mastering these skills isn't just about closing deals; it's about creating connections, solving problems, and transforming lives.

Next, let's explore how AI is reshaping the sales landscape and why these timeless skills are more relevant than ever in an AI-driven world.

CHAPTER 4

Negotiation Mastery

At the heart of every successful deal lies one critical skill: negotiation. Negotiation isn't just about haggling over price—it's the art of creating mutually beneficial agreements that satisfy all involved parties.

Negotiation is the process of reaching an agreement between two or more parties who have different needs, priorities, or perspectives. It's not about winning or losing—it's about finding a solution where everyone feels they've gained something valuable.

Master negotiators don't just close deals; they build partnerships, forge trust, and unlock opportunities that transform companies and industries. From the boardroom to the sales floor, negotiation has shaped empires, and mastering this skill can unlock unparalleled success.

Further, negotiation isn't about overpowering the other side; it's about collaboration. The best negotiators approach discussions with empathy, patience, and a genuine desire to find common ground.

There are four key principles of negotiation:

Preparation is Everything

Great negotiators spend more time preparing than negotiating. They understand their goals, the other party's goals, and the context of the deal.

Know Your BATNA (Best Alternative to a Negotiated Agreement)

Your BATNA is your fallback option if negotiations fail. Knowing it strengthens your position and ensures you don't accept a bad deal.

Listen Before Speaking

A critical reminder: Understanding the other party's needs and priorities is of utmost importance. The more you know, the better you can craft an offer that works for both sides.

Focus on Interests, Not Positions

Instead of arguing over your customer's surface-level demands, dig deeper to understand the underlying needs driving those demands. Remember, *hear* your customer, don't simply listen.

THE PSYCHOLOGY
OF NEGOTIATION

Negotiation is more than a sales tactic. It's a life skill. Whether you're closing a deal, securing a promotion, or resolving a conflict, negotiation empowers you, as a salesperson, to achieve your goals while strengthening relationships. Mastering negotiation isn't just about getting what you want; it's about creating value, fostering collaboration, and building a reputation as someone who delivers results.

Negotiation is as much about psychology as it is about strategy. People agree to deals not just because they make logical sense but because they feel understood, respected, and valued. Negotiation has shaped some of the most successful companies and industries in history. From forging groundbreaking deals to resolving conflicts, it's a skill that separates the average from the exceptional.

There are three primary psychological tactics utilized in negotiation. *Anchoring* is first, as it sets the tone. By starting with an ambitious but reasonable offer, you can influence the range of the discussion. Next is *reciprocity*. People feel compelled to give something in return when they've received something first. Small concessions can encourage the other party to reciprocate. Finally, there's *framing*. The way you present information matters. Framing an offer as a gain rather than a loss can make it more appealing.

While not one of the three primary tactics, emotions also play a powerful role in negotiation. It's important to stay calm and composed, but recognizing and managing emotions—both yours and the other party's—can help you navigate difficult discussions and reach better outcomes.

Primary Negotiation Tactics
Used in Negotiation

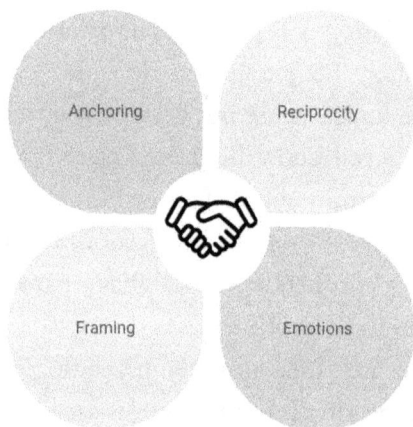

Anchoring

Reciprocity

Framing

Emotions

Negotiation isn't limited to contracts and deals. It shapes entire industries. Here are some examples of how negotiation mastery transformed companies.

CASE STUDY 1

Steve Jobs and the Music Industry

When Steve Jobs launched iTunes, he had to convince record labels to sell their music digitally at a fixed price of $0.99 per song. The music industry was resistant, fearing it would undermine album sales. Jobs used several negotiation tactics.

Vision building allowed him to paint a picture of how digital music could grow the industry. Through *creative leveraging*, Apple's brand and its growing iPod sales gave Jobs a strong bargaining position. And with *patience*, Jobs persisted, revisiting discussions until he got buy-in.

The result? iTunes revolutionized the music industry and became one of Apple's most profitable ventures.

CASE STUDY 2
Tesla and Supply Chain Negotiations

Tesla's success is partly due to its ability to negotiate favorable deals with suppliers. For example, Tesla secured long-term contracts for lithium-ion batteries by offering volume commitments. The promise of large orders gave suppliers confidence in steady revenue.

They also positioned themselves as a partner in

innovation, appealing to suppliers' desire to be part of a groundbreaking company.

These negotiations allowed Tesla to scale production and maintain a competitive edge in the electric vehicle market.

5 SUCCESSFUL NEGOTIATION TECHNIQUES

Stay Quiet

In negotiations, silence is golden. When you pause after making an offer or counteroffer, it puts pressure on the other party to respond. Silence also shows confidence and gives you the upper hand.

Ask Open-Ended Questions

Instead of asking, "Will you accept this price?" ask, "What concerns do you have about this price?" Open-ended questions encourage dialogue and help uncover valuable information.

Create Win-Win Scenarios

Find solutions that address both parties' priorities. For example, if a customer is concerned about price, offer flexible payment terms instead of a discount.

Build Urgency

Use time constraints to encourage decisions. For instance, saying, "This promotion is available until the end of the month" creates a sense of urgency without applying undue pressure.

Know When to Walk Away

Sometimes, the best deal is no deal. If a negotiation isn't meeting your minimum requirements, be prepared to walk away. Confidence in your position often leads the other party to reconsider their stance.

CASE STUDY
Disney's Acquisition of Marvel

Disney's acquisition of Marvel for $4 billion in 2009 was a masterclass in negotiation. Disney recognized Marvel's value as a brand but had to navigate several challenges, including securing rights to Marvel's intellectual property. The deal required:

Creative Structuring

Disney agreed to give Marvel's top executives creative control over future films, addressing their concerns.

Long-Term Vision

Disney communicated its plan to elevate Marvel's characters into a global franchise. The acquisition turned out to be a monumental success, with Marvel becoming one of Disney's most profitable divisions.

AI enhances negotiation in three primary ways. *Data-driven insights* are possible, given the way AI can analyze past deals to identify patterns and suggest optimal strategies. There's also the value of *predictive analysis*; AI tools can forecast the likelihood of different outcomes, helping you prepare for counteroffers. And let's not forget *efficiency*. Automated systems can handle routine negotiations, freeing up time for high-value discussions.

As AI tools become more prevalent, they're transforming the way we approach negotiation. AI can analyze data, predict outcomes, and provide real-time insights to strengthen one's position. However, as I've already mentioned (and will continue to, as it's critically important), successful negotiations will always require the human touch.

As the sales landscape evolves to incorporate AI and technology, a salesperson's negotiation skills will remain a critical differentiator. The ability to understand human needs, manage emotions, and create win-win outcomes is irreplaceable, even in an AI-driven world. As we dive into the future, remember this: Negotiation is timeless, and those who master it will always have an edge.

PART 2

Why the Sales Professional is Still Relevant

In an era dominated by artificial intelligence and automation, it's easy to question the future of the sales professional. After all, if AI can analyze data, personalize outreach, and close deals with greater efficiency, where does that leave the human salesperson?

The truth is, while AI is transforming sales, it cannot replace the emotional intelligence, creativity, and relationship-building skills that define great sales professionals. Salespeople are not just deal-closers; they're trusted advisors, problem solvers, and, most importantly, the human touch behind every transaction.

CHAPTER 5

The Case for Sales Expertise in an AI World

We live in an age when artificial intelligence can write emails, respond to customer inquiries, and even close deals. The temptation is, therefore, to assume that the traditional role of the sales professional is on the verge of extinction. After all, if technology can outperform humans in speed and efficiency, why would salespeople still be needed?

The truth is, AI excels in areas where logic and data are king, but it cannot replicate the depth of human involvement in business. Not only will sales professionals remain critical, even in an AI-driven world, but their unique skills will continue to complement and elevate AI to create better outcomes.

THE RISE OF AI IN SALES
WHAT IT DOES WELL

Artificial intelligence has fundamentally re-shaped the sales process by automating repetitive tasks and analyzing large datasets at speeds humans cannot match. AI has several key strengths, and it's important to be clear on what those are.

First, there's the area of *lead scoring*. AI can analyze historical data to identify which leads are most likely to convert, allowing sales teams to focus their energy on high-value prospects. Second is the area of *customer insights*. AI can track customer behavior, preferences, and purchase history to generate personalized recommendations. Finally, there's the area of *automated outreach*, which you're no doubt already quite familiar with, at least as a consumer. This includes the chatbots, email automation, and SMS campaigns powered by AI that are able to engage with hundreds of leads simultaneously.

These capabilities make AI an invaluable tool for scaling sales operations, improving efficiency, and reducing costs. Yet despite its strengths, AI has clear limitations.

For one, AI lacks emotional intelligence. As such, it can't pick up on subtle emotional cues like hesitation, excitement, or doubt. It also lacks creativity. While it can follow patterns and rules, it isn't able to develop innovative, out-of-the-box solutions to complex problems. And then there's that oft-mentioned trust-building component. There's simply no denying that customers are more likely to trust a person they can relate to than a machine, especially when making high-stakes decisions. So, while AI can handle many transactional aspects of sales, it lacks the depth and nuance required to foster the degree of genuine human connection necessary for successful sales transactions.

Sales is, ultimately, about people. Even in a world dominated by automation, human relationships remain central to closing deals, especially in high-value transactions. Customers want to feel heard, understood, and valued, and these are qualities that only humans can deliver.

Consider this: Would you trust an AI realtor to negotiate the purchase of your dream home or an AI financial advisor to guide you through a complex investment decision? For many, the answer is no. The reassurance of a knowledgeable (human) professional is irreplaceable in scenarios where emotions and stakes run high.

Experienced sales professionals also excel in reading between the lines. They can detect a customer's hesitation through tone of voice, body language, or subtle pauses—things AI cannot interpret effectively (at least, not yet). Empathy allows salespeople to adapt their approach, address concerns, and build trust in ways that feel authentic.

For example, a customer may express concerns about the cost of a product. While AI might respond by offering a sterile discount, a skilled salesperson can dig deeper to uncover the root of the customer's concern and offer a tailored solution that addresses the customer's true needs. In so doing, they're more likely to close the deal.

Of course, sales isn't just about closing deals—it's about fostering relationships that lead to repeat business and referrals. Customers are loyal to people, not machines. Sales professionals who invest in building trust and delivering value become invaluable partners to their clients.

THE HYBRID MODEL HOW HUMANS AND AI CAN WORK TOGETHER

The future of sales doesn't require choosing between humans and artificial intelligence; it's about combining the strengths of both. AI can handle the heavy lifting—data analysis, lead generation, and routine follow-ups—allowing sales professionals to focus on what they do best: engaging with their customers and closing deals. AI can analyze hundreds of leads to identify the top ten worth pursuing, while a skilled salesperson can then take those top leads and use their expertise to close deals.

One of AI's most powerful contributions is its ability to personalize customer experiences. By analyzing data, it can identify a customer's preferences, habits, and pain points. Armed with this information, sales professionals can proactively tailor their approach to resonate with the individual.

For instance, an AI-powered CRM might alert a car salesperson that a repeat customer prefers SUVs and values fuel efficiency. The salesperson can then use this insight to recommend the perfect vehicle, strengthening the relationship and increasing the likelihood of a sale.

Sales professionals provide unique types of value. In the area of problem-solving and creativity, sales professionals thrive in situations where creativity and adaptability are required. While AI follows pre-programmed rules, humans can think on their feet and come up with innovative solutions to complex problems.

Consider a business negotiation where unexpected challenges arise, such as budget constraints or shifting timelines. A skilled salesperson can navigate these hurdles with creativity and flexibility, finding a solution that satisfies all parties.

Persuasion is an art that AI simply cannot master. Closing a deal often requires understanding a customer's motivations, addressing their fears, and inspiring confidence. This is where sales professionals can successfully use storytelling, emotional appeal, and charisma to influence decision-making in ways that go beyond logic. A salesperson might, for example, share a story about how their product transformed another client's business, creating an emotional connection that convinces the customer to move forward.

I'm sure it will come as no surprise that while AI can provide information, it can't offer authenticity. Customers are more likely to trust a person who shares their values, listens to their concerns, and demonstrates genuine care. Again, sales professionals have the unique ability to humanize the sales process, making customers feel valued rather than treated as just another data point.

CASE STUDY 1
Salesforce and AI Integration

Salesforce's AI tool, Einstein, helps sales teams by analyzing customer data and providing actionable insights. However, the sales professionals using Einstein remain the ones building relationships, negotiating deals, and closing sales. This hybrid approach has led to significant increases in productivity and revenue.

CASE STUDY 2
Real Estate Sales

In the real estate industry, AI tools can provide market trends, property recommendations, and pricing insights. Yet buyers still rely on real estate agents to guide them through the emotional and financial complexities of purchasing a home. The agent's ability to empathize, negotiate, and provide reassurance is irreplaceable.

PREPARING FOR THE FUTURE OF SALES

The sales professionals who thrive in an AI-driven world are those who commit to continuous growth. By staying updated on industry trends, learning to use AI tools effectively, and refining their core sales skills, they will position themselves as indispensable.

As AI begins to handle routine tasks, sales professionals must shift toward more of a consultative role. This means they must begin to focus on understanding their customer's unique goals, offering strategic advice, building trust, and delivering long-term value. In a world of automation, relationships are the ultimate competitive advantage. The future of sales belongs to those who can combine the power of technology with the timeless skills of selling.

Up next, we'll explore how mastering these skills can elevate a salesperson from an amateur to a true professional, ensuring they remain indispensable in an AI-driven world.

CHAPTER 6

From Amateur to Master

The Value of Learning Your Craft

E very profession has both amateurs and masters. Amateurs dabble, rely on luck, and often plateau early. Masters, on the other hand, approach their craft with discipline, intentionality, and a commitment to continuous improvement. Sales is no different. The gap between an average salesperson and a masterful one is seen not merely in results; it's also in how they approach the art and science of selling.

In a world increasingly influenced by AI and automation, the demand for highly skilled sales professionals has never been greater. AI can handle the basics, but mastery is required to elevate sales from transactions to transformative experiences. In this chapter, we'll explore what it takes to evolve from an amateur to a master salesperson, the skills required, and why learning the craft is more important than ever.

Many people mistakenly believe that sales is only about being persuasive or having the ability to "convince" someone to buy something they don't need. In reality, sales is about understanding, solving problems, and creating value for the customer. And when done right, it's one of the most respected and rewarding professions in the world.

Sales has come a long way from the stereotype of the slick-talking salesperson. Today, it's a highly specialized field requiring technical expertise, emotional intelligence, and on-the-fly strategic thinking. Companies rely on sales professionals not just to close deals but to drive growth, build relationships, and shape customer experiences.

THE AI ERA
WHY HUMAN SKILLS STILL MATTER

AI can help sales teams by automating repetitive tasks, analyzing customer data, and even making recommendations. But still, the human element can't be replaced. In fact, as AI takes over the "easy" parts of sales, the demand for skilled professionals to handle the more challenging parts will only grow.

Of course, achieving mastery in sales doesn't happen overnight. It requires dedication, practice, and a willingness to learn from one's own as well as others' successes and failures. Even the best sales professionals never outgrow the basics. Core skills like active listening, effective communication, and understanding customer needs are the foundation of success. Without these fundamentals, no number of advanced techniques or AI tools will make a difference.

Emotional intelligence (EQ) is the ability to understand and manage one's own emotions while also recognizing and influencing the emotions of others. In sales, EQ is critical for building trust and rapport with customers; reading subtle cues, like hesitation or enthusiasm; managing rejection; and staying motivated.

Amateurs stop practicing once they feel "good enough." Masters, on the other hand, understand that there's always room to improve. Whether it's through role-playing, reviewing recorded sales calls, or studying customer feedback, consistent practice is what separates the good from the great.

Don't ever take for granted that every failed sale is an opportunity to learn. Masters take the time to analyze what went wrong after each failed transaction, whether it was a missed buying signal, an objection that wasn't handled properly, or a lack of preparation. They view mistakes not as failures but as steppingstones to success.

The fastest way to grow in sales is by learning from those who've already mastered the craft.

Whether it's a seasoned colleague, a manager, or a professional coach, having a mentor provides guidance, accountability, and a fresh perspective.

CORE SKILLS OF A MASTER SALESPERSON

When it comes to the skills a master salesperson has honed, five continue to stand out.

Active Listening

The best salespeople listen more than they speak. Active listening involves focusing fully on the customer, rather than thinking about what you'll say next; asking clarifying questions to ensure you understand their needs; and reflecting back what you've heard to show you're paying attention.

Storytelling

Storytelling turns features into benefits and makes a pitch memorable. Instead of listing product specs, master salespeople share stories that demonstrate how their product or service solves real-world problems.

Objection Handling

Every customer has objections, whether about price, timing, or perceived value. Masters don't fear objections—in fact, they welcome them as opportunities to address concerns and build trust.

Negotiation

Negotiation is about creating a win-win scenario where both parties feel they've gained something valuable. Masters are skilled at balancing assertiveness with collaboration, ensuring the deal benefits everyone involved.

Adaptability

No two customers are the same, nor are any two sales situations. Master salespeople know how to read the room, adapt their approach, and pivot when necessary.

CORE SKILLS OF A MASTER SALESPERSON

1 Active Listening

2 Storytelling

3 Objection Handling

4 Negotiation

5 Adaptability

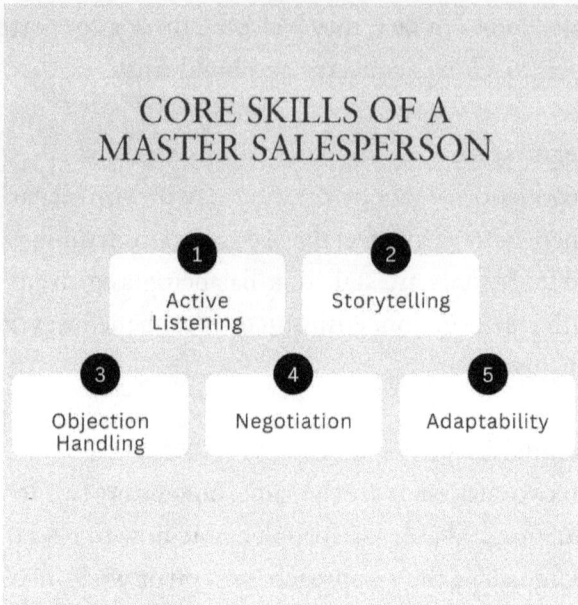

Transactional selling (which completely leaves out the critical human element) focuses solely on closing the deal as quickly as possible. It's short-term, impersonal, and often focused on price rather than value. While it may work in certain industries, it rarely leads to long-term success.

Master salespeople embrace the concept of consultative selling, which is centered on understanding the customer's needs and offering tailored solutions. This approach builds trust, fosters loyalty, and often leads to higher-value deals. Key elements of consultative selling include asking probing questions, which allows a salesperson to uncover the customer's deeper goals and challenges. Additionally, master salespeople provide expert advice. They position themselves as a trusted advisor rather than just a vendor. Finally, masters of sales focus on the relationship. They prioritize long-term partnerships over immediate sales.

CASE STUDY
The Consultative Approach in Action

A software salesperson encounters a client who is hesitant to adopt a new platform. Instead of pushing the product, they ask about the client's current workflow challenges. Through careful questioning, they're able to discover that the client's team struggles with collaboration. In response, the salesperson then highlights how the software's collaboration tools can address this pain point. The result? A deal closed with full buy-in from the client.

THE ROLE OF CONTINUOUS LEARNING

The sales landscape is constantly evolving. New technologies, shifting customer expectations, and changing market dynamics mean that what worked yesterday might not work tomorrow. Masters stay ahead by continually learning and adapting.

Sales professionals who commit to learning will always have a competitive edge. They're not just keeping up with changes, they're leading the charge. They schedule in activities such as reading books; blogs; and articles on sales, psychology, and business can provide fresh insights and ideas. They also frequently attend workshops and conferences. Engaging with other professionals and industry leaders helps them stay updated on the latest trends. They embrace technology, learning how to use AI tools, CRMs, and data analytics to enhance their performance. And, perhaps most important, they reflect regularly, taking time to review their successes and failures as well as identifying areas for improvement.

As AI takes over routine tasks, uniquely human skills like creativity, emotional intelligence, and relationship-building become even more valuable, and master salespeople who excel in these areas will stand out in an increasingly automated world.

There are a number of rewards that come with AI mastery, including higher earnings, career longevity, and personal fulfillment. Master salespeople consistently outperform their peers and earn higher commissions. As industries evolve, those with advanced skills remain in demand. And mastery brings confidence, satisfaction, and pride in one's work.

Becoming a master salesperson isn't about talent—it's about *commitment*.

It's about showing up every day with a desire to

learn, grow, and serve your customers better. In an AI-driven world, the mastery of sales skills will be the ultimate differentiator. Those who invest in their craft, embrace lifelong learning, and focus on creating value will not only thrive but they'll also redefine what it means to be a sales professional.

Next, we'll explore the emotional intelligence and human connection that make sales professionals indispensable, even as technology continues to advance. Mastery starts with understanding, and it ends with transforming how we sell in a rapidly evolving world. So let's dive deeper.

CHAPTER 7

The Human Element
Building Trust in a Digital Age

In sales, trust is everything. It's the foundation of every relationship, the fuel for long-term loyalty, and the deciding factor when customers face a choice between you and your competitors. In a world dominated by AI, automation, and endless streams of digital communication, trust has become even more valuable. Customers are constantly inundated with information, and while AI can offer efficiency and personalization, it can't replicate the authenticity and empathy that come from human interaction.

The rise of digital tools has transformed how businesses connect with customers. AI chatbots, automated emails, and targeted ads make reaching potential buyers easier than ever. However, this constant bombardment of messages has led to customer fatigue and skepticism. Many buyers are wary of overly polished, impersonal sales tactics and crave genuine, human connection.

These days, customers are more cautious, doing extensive research before making purchasing decisions. They want to feel confident that they're making the right choice, and that confidence often comes from interacting with a trustworthy, knowledgeable salesperson. Trust acts as a shortcut in decision-making. When customers trust a salesperson, they're more likely to share their true concerns and needs, accept recommendations without second-guessing them, and forgive minor mistakes or mishaps.

Building trust isn't just a nice-to-have—it's a must-have for long-term success in sales.

HOW SALES PROFESSIONALS BUILD TRUST

Authenticity is the cornerstone of trust, and customers can often tell when a salesperson is being insincere or overly rehearsed. Instead of trying to appear flawless, it's important to embrace honesty and transparency. We can do this in several ways:

Admit Limitations
If your product isn't the best fit for a particular customer's needs, outright acknowledge it. Customers appreciate honesty over forced persuasion.

Share Personal Stories
Relating to customers through personal anecdotes makes you more relatable and approachable.

Avoid Overpromising
Set realistic expectations and then be sure to deliver on them.

Having the quality of empathy allows you to understand and share the feelings of others. In sales, empathy means genuinely caring about the customer's challenges and goals, not just about making the sale.

Be sure to ask thoughtful questions. Go beyond surface-level inquiries to uncover deeper needs. Validate your customer's concerns. Show your customers that you understand their hesitations, and address them with patience. Finally, do your best to put yourself in your customer's shoes. Imagine how you would feel in their position, and allow that to help guide your responses.

Trust isn't built in a single interaction; it's the result of consistent behavior over time. That's why it's critical to follow through you on promises—if you say you'll do something, do it, whether it's sending additional information or following up at a specific time. Be reliable. Show up prepared for meetings, respond promptly, and stay accessible.

Further, don't neglect the importance of maintaining regular communication with your customers. Keep them updated, even if there's no major news to share.

THE SCIENCE OF TRUST IN SALES

Research shows that trust activates specific areas of the brain linked to reward and decision-making. When customers trust you, they feel safe and are more likely to make positive decisions. Conversely, distrust triggers fear and caution, making customers hesitant to move forward.

According to the "Trust Equation," trust is built on three key factors: credibility, reliability, and intimacy. Credibility relies on the expertise and knowledge you have to guide the customer. Reliability depends upon whether or not you consistently deliver on your promises. And intimacy is fostered when you create a sense of connection and understanding with your customer. When these factors are present, trust flourishes. When they're missing, trust wanes if not dissolves.

THE TRUST EQUATION
3 key factors

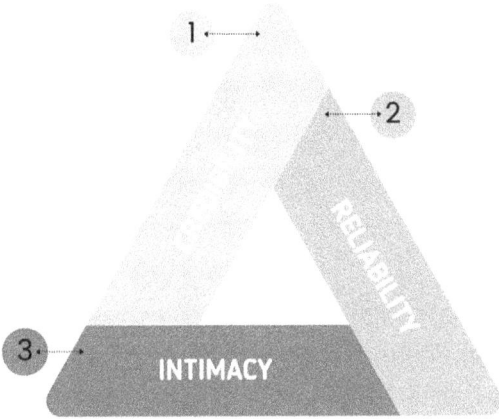

CASE STUDY 1
The Trusted Advisor

A financial advisor working with a hesitant client on retirement planning prioritizes understanding their fears about market volatility. By asking empathetic questions and explaining options without pushing for an immediate decision, the advisor builds trust over multiple meetings. The result? A loyal client who not only invests but also refers friends and family.

CASE STUDY 2
Recovering from Mistakes

A car salesperson accidentally quotes the wrong price for a trade-in during negotiations. Instead of avoiding the issue, they immediately own up to the error, apologize sincerely, and offer a fair resolution. The customer appreciates the honesty and proceeds with the purchase, confident in the salesperson's integrity.

While AI can't replace human trust, it can certainly support the trust-building process. AI tools can analyze customer data to create tailored recommendations, making interactions feel more relevant. Automation also makes a salesperson's activities more efficient by ensuring timely follow-ups, reducing the chances of a customer feeling ignored. Finally, AI can provide clear, data-backed insights to reinforce the salesperson's credibility.

The key to building trust in the digital age is knowing when to rely on technology and when to step in as a human. Use AI to handle initial outreach and provide insights, but be sure to step in as a human to address concerns, answer complex questions, and build rapport.

CREATING A SEAMLESS
CUSTOMER EXPERIENCE

In today's world, customers are more skeptical than ever, especially if they've been burned by previous sales experiences. Patience, transparency, and empathy are therefore essential for overcoming this skepticism. Customers should never feel like they're "just another number." Even in digital interactions, adding a personal touch—such as mentioning specific details from past conversations—can strengthen trust.

Trust isn't just about closing a single deal—it's about fostering relationships that lead to repeat business and referrals. Loyal customers are your greatest advocates, and they provide long-term value far beyond an initial transaction.

When customers trust you, they're more likely to recommend you to others. They're also less price-sensitive, focusing instead on the value you deliver. And they actually become partners in your success, offering feedback and insights that help you improve. In industries where products and prices are similar, trust becomes the ultimate differentiator. Customers will choose the salesperson they trust, even if the competitor offers a lower price.

Mistakes are inevitable, but they don't have to destroy trust. Address them head-on, apologize sincerely, and offer a solution. Customers will often respect you more for how you handle errors than if everything goes perfectly.

Sales quotas and tight deadlines often tempt sales professionals to take shortcuts or overpromise. However, compromising authenticity for short-term gains will erode trust in the long run, which is a problem because trust is the most valuable currency in sales. It's what transforms interactions into relationships, customers into advocates, and salespeople into trusted advisors.

In a digital age filled with noise, the sales professionals who prioritize trust will stand out. By mastering the art of building and maintaining trust, you not only secure immediate wins but also lay the foundation for long-term success.

CHAPTER 8

Thriving in the AI-Driven Sales Landscape

The sales landscape is evolving rapidly, and professionals who adapt will flourish. Embracing AI doesn't mean abandoning traditional sales techniques—it means understanding how to use technology to enhance them.

Three core skills set master sales professionals apart from their inexperienced counterparts. First, their *tech savviness*, or understanding of how to use AI tools effectively to streamline processes. Second is *personalization*, or an ability to deliver customized experiences that AI alone cannot replicate. Finally, there's *thought leadership*: positioning yourself as a trusted advisor by staying informed and sharing valuable insights with customers.

THE FUTURE ROLE OF
SALES PROFESSIONALS

As AI takes over routine tasks, the role of the sales professional will shift toward more strategic and consultative activities, and professionals who can combine their human expertise with AI insights will become indispensable to their organizations.

There are several key ways that sales professionals can stay relevant in this evolving environment:

Embrace Lifelong Learning
Stay updated on industry trends and technological advancements.

Focus on Relationships
Prioritize building trust and connections with customers.

Leverage AI Tools
Use AI to handle administrative tasks, freeing up time for high-value interactions.

Yes, sales is evolving, but the core principles remain the same, and sales professionals who master the arts of listening, storytelling, and relationship-building will thrive in the changing landscape. The key is to view artificial intelligence as an ally, not a threat. By combining the timeless skills of selling with the power of technology, sales professionals can achieve unprecedented levels of success. In the next section, we'll explore how to capitalize on emerging AI platforms to amplify your sales efforts and stay ahead of the curve.

The future of sales isn't about choosing between humans and machines—it's about combining the strengths of both to create extraordinary results.

PART 3

AI's Takeover and Transformation of Sales

Artificial intelligence has redefined the sales industry, introducing unprecedented efficiency, scale, and precision. AI tools can analyze data, automate outreach, and even close deals faster than most humans can.

At the same time, AI is only as powerful as the data that drives it. Without clean, accurate, and organized data, even the most advanced AI tools fall short. As Larry Ellison once said, "Unless you have your data organized, you can't use AI." This statement underscores a critical reality:

The foundation of AI-driven sales success is high-quality data.

In this section, we'll explore how AI is transforming the sales process, the importance of having the right data, and how the newest AI tools can bridge the gap by ensuring your data is clean, complete, and ready to power your sales strategy.

CHAPTER 9

AI in Sales
The Game Changer

Artificial Intelligence has fundamentally re-shaped the sales industry, revolutionizing how businesses engage with customers, manage leads, and close deals. What once relied heavily on human intuition and labor-intensive processes is now augmented by advanced algorithms that analyze data, automate repetitive tasks, and provide insights at unprecedented speed and scale.

AI isn't just another tool in the sales toolbox; it's a paradigm shift. While it streamlines and enhances many aspects of sales, it also creates new opportunities for sales professionals to focus on high-value activities that require human touch, creativity, and emotional intelligence.

AI excels at analyzing vast amounts of data to identify high-quality leads. By assessing customer behaviors, demographics, and purchase history, AI can predict which leads are most likely to convert. This enables sales teams to focus their efforts on prospects with the highest potential.

Gone are the days of generic email blasts. AI now enables salespeople to create specific, hyper-personalized communication by analyzing customer preferences, behaviors, and past interactions. Whether it's crafting tailored email subject lines or suggesting personalized offers, AI ensures that every message resonates with the recipient. AI-powered platforms such as HubSpot or Outreach.io can generate custom email campaigns based on a customer's recent interactions with your brand. And AI chatbots can engage prospects in real-time on your website, answering questions and guiding them toward a purchase.

AI tools can also track website activity, like page views or time spent on specific pages, to identify leads that show strong buying intent. Further, lead scoring systems powered by AI can assign numeric values to leads based on their likelihood to convert, ensuring that sales representatives prioritize their outreach effectively.

Accurate sales forecasting is critical for effective planning and resource allocation. AI analyzes histor-ical sales data, market trends, and external factors to predict future sales performance with remarkable ac-curacy. Additionally, AI tools can predict seasonal fluctuations in demand and recommend strategies to capitalize on upcoming opportunities. Plus, forecasting models can alert managers to potential shortfalls in pipeline coverage, enabling proactive adjustments.

AI ensures that no lead falls through the cracks by automating follow-up communication. Whether it's a reminder email, a thank-you note, or a call scheduling prompt, AI-powered tools can handle these routine tasks, freeing up sales reps to focus on more complex interactions.

> "Sales AI is making it easier and better to work, but not by taking jobs from sales reps."
>
> —*Cory Benz, Revenue Operations Manager at Crexi*

As an example, tools like Salesforce Einstein or Gong.io can automatically remind sales reps when it's time to follow up with a lead or customer, and AI systems can trigger personalized messages based on customer behavior, such as abandoned shopping carts or webinar sign-ups.

THE BENEFITS OF AI IN SALES

AI offers numerous benefits that make it indispensable for modern sales organizations. For one, AI automates time-consuming tasks like data entry, lead scoring, and follow-ups, allowing sales teams to focus on building relationships and closing deals. This increased efficiency translates into higher productivity and lower operational costs.

AI also provides actionable insights based on data analysis, helping sales teams make smarter decisions. By identifying trends, patterns, and opportunities, AI empowers sales professionals to approach every interaction with confidence and precision. Further, AI allows businesses to scale their sales operations without proportional increases in headcount. Automated tools can handle large volumes of leads and customers simultaneously, ensuring consistent engagement at every stage of the funnel.

With AI-driven personalization, customers receive relevant and timely interactions that address their specific needs. This creates a seamless and engaging experience, building trust and loyalty.

But remember, AI is only as effective as the data it relies on. Clean, accurate, and organized data is the lifeblood of AI systems. Without it, even the most advanced AI tools will deliver subpar results.

THE IMPORTANCE OF
DATA HYGIENE

Dirty data—which is characterized by missing information, duplicates, and inaccuracies—can lead to ineffective lead scoring, poor customer targeting, and wasted time and resources.

To address the challenge of dirty data, platforms like Spearphish.io offer a comprehensive data hygiene process. By integrating with your CRM, Spearphish.io can clean your data, removing duplicates, correcting errors, and standardizing formatting. It can also append missing information, adding crucial details like phone numbers, email addresses, and physical addresses. It's able to verify accuracy, ensuring your data is up-to-date and reliable.

Clean data enables sales teams to spend less time correcting errors or chasing dead leads, engage with high-quality prospects more effectively, and deliver personalized experiences that drive conversions. With clean and complete data, AI systems can deliver accurate insights, personalized outreach, and improved overall performance.

CASE STUDY
Data Hygiene in Action

To illustrate how clean data fuels AI and drives tangible sales results, take the example of a B2B software company that struggled with outdated and incomplete CRM data, leading to low email open rates and poor lead conversion. After using AI software to clean and append their data, email open rates increased by 40%, lead-to-opportunity conversion rates improved by 30%, and sales reps reported spending less time fixing data and more time engaging with qualified prospects.

THE HYBRID MODEL
AI AND HUMANS WORKING TOGETHER

The future of sales isn't about choosing between humans and AI; it's about combining their strengths to create a powerful partnership.

AI takes care of the heavy lifting—data analysis, task automation, and trend prediction—allowing sales professionals to focus on high-value activities like building relationships, solving complex problems, and closing strategic deals.

As AI automates routine tasks, the role of the sales professional will shift to become more consultative. Salespeople will act as trusted advisors, guiding customers through high-stakes decisions and delivering tailored solutions.

AI has undeniably transformed the sales industry, offering unprecedented efficiency, scalability, and personalization. However, its success depends on having clean, accurate data and a team of skilled professionals who know how to leverage its capabilities.

By combining the power of AI with the creativity, empathy, and expertise of human sales professionals, businesses can achieve extraordinary results. AI isn't just changing how we sell—it's redefining what's possible in sales. The game has changed, and those who embrace this transformation will thrive in the new era of sales.

CHAPTER 10

The Role of Data in AI-Driven Sales

Artificial intelligence has revolutionized sales, offering unprecedented levels of automation, personalization, and insight. However, as I've mentioned, the effectiveness of AI is directly tied to the quality of the data it relies on. Without clean, accurate, well-structured data, even the most advanced AI systems will fail to deliver meaningful results.

AI systems learn, predict, and make decisions based on the data they process. In sales, this means AI tools rely on data to identify high-potential leads, analyze customer behavior and preferences, automate personalized outreach, and provide actionable insights to sales teams. Simply put, data is the fuel that powers AI. The better the data, the more effective the AI.

> *"Bots can be a salesperson's best friend when it comes to pre-qualifying leads and ensuring they have more time for higher-value tasks."*
>
> —*Mary Shea, Principal Analyst at Forrester*

At every stage of the sales funnel, data plays a critical role. When generating leads, data helps identify and qualify prospects based on key characteristics. In the engagement stage, insights from data enable personalized messaging that resonates with individual customers. When closing, data-driven insights guide sales teams on the best strategies to close deals. And in the area of retention, customer data informs loyalty programs and upsell opportunities. If the data feeding these processes is incomplete, outdated, or inaccurate, the entire sales operation suffers.

Dirty data refers to information that is inaccurate, incomplete, or inconsistent. According to industry studies, dirty data costs businesses an average of 15-25% of their revenue annually, and sales reps spend up to 30% of their time dealing with data quality issues instead of selling. The numbers make it clear: clean data isn't just a nice-to-have—it's a business necessity. Common dirty data issues include:

Missing Contact Details
Phone numbers, emails, or addresses are often

incomplete or outdated.

Duplicate Records
Multiple entries for the same customer create confusion and inefficiency.

Inconsistent Formatting
Variations in how data is recorded (e.g., "NY" versus "New York") hinder analysis.

Outdated Information
Customer preferences or contact details may no longer be valid.

Poor data hygiene has real consequences for sales teams. These include:

Wasted Time
Sales reps spend valuable hours correcting data instead of engaging with customers.

Lost Opportunities
Inaccurate or incomplete data leads to missed chances to connect with high-potential leads.

Ineffective AI
Dirty data undermines the performance of AI tools, resulting in inaccurate predictions and poor recommendations.

Damaged Reputation

Incorrect or repetitive outreach can frustrate customers and harm your brand image.

SOLVING THE DATA PROBLEM

To succeed in an AI-driven sales landscape, businesses must adopt a data-first mindset. This means prioritizing data accuracy (ensuring that all information is correct and up to date), data completeness, filling in gaps to create a comprehensive customer profile, and data consistency (standardizing how data is recorded and stored).

Clean data allows AI tools to provide accurate insights and recommendations. It also enables sales reps to spend less time fixing data and more time engaging with qualified leads. Accurate data ensures personalized and relevant interactions with customers. And with better insights and more accurate targeting, sales teams can close more deals.

Clean data doesn't just improve AI performance; it creates a feedback loop that enhances your entire sales operations. It powers AI tools, AI tools generate insights and automate processes, these insights drive more effective sales strategies, and improved strategies generate better data for AI analysis. This cycle ensures continuous improvement, but it all starts with clean data.

In support of this goal, Spearphish.io offers three key services. First, *data cleaning*. The platform removes duplicates, corrects errors, and standardizes formatting to ensure consistency. Next is *data appending*. Missing information, such as phone numbers, email addresses, and physical addresses, is added to your records. And finally, *data verification*, where Spearphish.io validates the accuracy of your data, ensuring it's up-to-date and reliable.

REAL-WORLD EXAMPLE

A financial services company struggled with low email open rates and poor lead engagement. After implementing AI to clean and append their data, email open rates increased from 18% to 45%, lead conversion rates improved by 35%, and sales teams reported spending 20% more time engaging with qualified prospects. This transformation highlights the importance of clean data in unlocking AI's full potential.

Data hygiene most certainly is not a one-time effort; it's an ongoing process. Regular audits and updates are essential in order to maintain data quality and ensure that your AI tools continue to deliver results.

Even with tools like Spearphish.io, sales professionals play a critical role in maintaining data quality. They must input data accurately and consistently, update customer records after every interaction, and flag issues or inaccuracies for correction. When sales teams and AI systems work together to prioritize data hygiene, the results are transformative.

> "True collaboration is not about dividing work between machines and people but about bringing the strengths of both together to solve problems and achieve more than either could alone."
>
> —*Garry Kasparov, Chess Grandmaster and Writer*

CHAPTER 11

From Zero to Guru
How AI Levels the Playing Field

For decades, sales success has often been determined by experience, intuition, and natural talent. Some sales professionals seemed to have an innate ability to close deals, while others struggled to break through. But with the rise of artificial intelligence, the sales landscape has shifted dramatically. AI has become the great equalizer, giving even the most inexperienced salespeople access to tools, insights, and strategies that were once the domain of top performers.

In this chapter, we'll explore how AI transforms novices into high-performing sales professionals, how it enhances skills at every level, and why it's reshaping the future of sales. We'll also highlight specific tools and techniques that sales teams can use to maximize their potential and achieve extraordinary results.

THE DEMOCRATIZATION OF SALES SUCCESS

AI empowers sales teams in a variety of ways. For one thing, it provides instant insights as AI tools analyze customer data to deliver actionable recommendations in real time, enabling even new sales reps to make informed decisions. It also provides guidance and coaching. AI platforms act as virtual mentors, offering feedback on calls, emails, and interactions to improve performance. Finally, there's efficiency at scale. Tasks that once required years of experience to master—like prioritizing leads or crafting personalized pitches—are now streamlined by AI.

This technology ensures that even those with limited experience can perform at a high level, allowing companies to build more balanced and effective teams.

CASE STUDY
The Newbie Advantage

A newly hired salesperson at a technology company used an AI-powered CRM to identify high-value leads and tailor their outreach. With no prior sales experience, they outperformed seasoned colleagues in their first quarter by relying on AI-driven insights and recommendations, demonstrating how AI can transform raw potential into measurable success.

One of the most powerful aspects of AI is its ability to provide personalized training. Tools like Gong.io and Chorus.ai analyze sales calls and emails to identify areas where reps can improve, such as tone, pacing, or objection handling. They also highlight successful strategies that can be replicated across the team and deliver real-time coaching tips during customer interactions. As an example, AI might suggest that a salesperson pause more frequently during a pitch or reframe a question to better address a customer's concerns.

AI doesn't just offer feedback after the fact, however. It also provides actionable recommendations in the moment. During a sales call, AI-powered platforms can suggest relevant product features based on the customer's questions, highlight cross-sell or upsell opportunities, and provide real-time prompts for handling objections. This real-time guidance enables salespeople to navigate complex conversations with confidence, even if they're new to the role.

Administrative tasks like data entry, scheduling follow-ups, and tracking progress can consume a significant portion of a salesperson's day. AI automates these tasks, freeing up time for reps to focus on what they do best: building relationships and closing deals. Platforms like Salesforce Einstein and

HubSpot use AI to handle routine tasks, ensuring that sales teams can maximize their productivity.

AI-POWERED TOOLS THAT DRIVE SUCCESS

AI analyzes historical data, customer behavior, and engagement metrics to rank leads based on their likelihood to convert. This ensures that sales teams focus their energy on the most promising opportunities. For example, Zoho CRM uses AI to prioritize leads by assigning scores based on customer activity, demographics, and previous interactions.

AI tools create hyper-personalized messages and campaigns tailored to each prospect's preferences and pain points. AI might suggest an email that references a recent product the prospect viewed online. Or, it could recommend a specific case study to share based on the prospect's industry. In fact, Outreach.io uses AI to craft email sequences that align with a lead's behavior and engagement history.

AI also enables predictive selling by analyzing data to anticipate customer needs. For example, if a customer's contract is about to expire, AI might recommend proactive outreach. AI can identify patterns, indicating when a lead is ready to buy. InsideSales.com, for example, provides predictive analytics that help sales teams know when and how to engage prospects.

OVERCOMING COMMON SALES CHALLENGES WITH AI

One of the most challenging aspects of sales is addressing objections. AI tools analyze past interactions to identify successful objection-handling strategies and suggest responses in real time. You've surely experienced a prospect mentioning budget concerns. In this instance, AI might prompt the rep to highlight ROI and offer flexible payment terms.

AI can also reduce the steep learning curve for new salespeople by acting as a virtual mentor or coach, providing guidance, suggesting next steps, and ensuring that sales reps are always prepared for customer interactions.

AI also enables sales managers to ensure consistent messaging and strategies across their teams by providing standardized recommendations, ensuring that all reps—regardless of experience—are aligned with the company's goals.

A mid-sized manufacturing company implemented an AI-driven sales platform to address declining close rates. Within three months, lead conversion rates improved by 25%, average deal size increased by 15%, and new sales reps were ramping up 40% faster, thanks to AI-driven training and support.

A SaaS company used AI to automate follow-ups and lead scoring. Sales reps reported spending 30% less time on administrative tasks and 20% more time engaging with qualified leads. The result? A 35% increase in revenue within six months.

As AI becomes more prevalent, the role of the salesperson is shifting. Sales professionals are no longer just deal-closers—they're trusted advisors who guide customers through complex decisions, and their ability to connect on a human level is what sets them apart in an AI-driven world. As a result, AI is reshaping the way sales teams are trained and developed, ensuring that sales teams are always improving and adapting to changing market conditions. There are now AI-powered platforms that use gamification to make training engaging and competitive while also identifying individual skill gaps and delivering tailored training content. Plus, AI provides instant feedback during practice sessions, accelerating skill development.

AI is also transforming sales by leveling the playing field, giving even inexperienced reps the tools and insights they need to succeed. By automating routine tasks, providing real-time guidance, and delivering actionable insights, AI empowers sales professionals to focus on what truly matters: building relationships, solving problems, and closing deals.

While AI enables efficiency and scale, it's the

combination of human expertise and AI-driven capabilities that creates true sales success. The future of sales belongs to those who embrace this partnership, leveraging AI to elevate their skills and deliver exceptional results.

CHAPTER 12

The Need for the Right Tools and Platforms

Artificial Intelligence has introduced unprecedented efficiency and scale into the sales process, but the real key to success lies in choosing the right tools and platforms to support this transformation. The right technology doesn't just enable AI—it amplifies its power, streamlines workflows, and ensures your sales team is equipped to thrive in the AI-driven era.

Let's explore why selecting the right AI tools is critical, which features to prioritize in AI sales platforms, and how technologies like Spearphish.io and IgniteUps.AI can empower teams to work smarter, not harder. We'll also discuss how leveraging these tools positions businesses to succeed in a rapidly evolving sales landscape.

THE RIGHT TOOLS MATTER

Sales is no longer just about making calls and sending emails—it's a complex process that involves analyzing customer data, automating outreach across multiple channels, personalizing interactions at scale, and monitoring performance and optimizing strategies.

AI tools are designed to manage this complexity, but without the right platforms in place, businesses risk missing opportunities, losing efficiency, and falling behind competitors.

Choosing the wrong tools, however, can hinder sales efforts in several ways. Overly complicated or poorly integrated systems slow down workflows. If a tool is difficult to use, your sales team won't fully adopt it, wasting both time and money. And tools without robust data analysis capabilities fail to provide actionable insights, leaving sales teams flying blind.

To truly capitalize on AI, businesses need platforms that align with their goals, integrate seamlessly with existing systems, and empower their teams to perform at their best.

When evaluating AI tools for sales, it's essential to focus on features that drive results. When it comes to the most important capabilities to look for, AI relies on clean, accurate data to perform effectively. Platforms such as Spearphish.io are invaluable because they integrate directly with your CRM to clean, append, and verify data; ensure your records are complete, consistent, and up to date; and eliminate duplicates and fill in missing contact information like phone numbers and email addresses. Remember, without high-quality data, even the best AI tools will deliver subpar results.

AI platforms should be capable of tailoring communication to each customer's unique preferences and behaviors. Look for tools that analyze customer data to create detailed profiles, craft personalized email campaigns, text messages, and call scripts, and suggest relevant products, services, or solutions during customer interactions.

For example, Outreach.io uses AI to personalize sales sequences, ensuring that every touchpoint resonates with the recipient. To support sales reps during live interactions, AI tools should provide real-time insights, such as suggested responses to objections; cross-sell and upsell opportunities based on customer data; and recommended next steps to move the deal forward. Gong.io, for instance, analyzes live calls and offers real-time coaching tips to improve outcomes.

AI tools should also reduce the time spent on administrative tasks, allowing sales teams to focus on building relationships and closing deals. Look for platforms that can automate follow-ups, reminders, and scheduling, handle lead scoring and routing, and streamline reporting and analytics. Take Hub-Spot, which automates routine sales tasks, enabling reps to spend more time engaging with customers.

One platform that's revolutionizing the sales process is IgniteUps.AI, which enables businesses to create AI-powered inbound and outbound sales agents. IgniteUps.AI is a game-changer in a number of ways. First, it allows businesses to deploy AI agents that can handle calls, emails, and text messages; engage with customers 24/7, providing instant responses and support; and execute tasks that would typically require a full call center staff. This functionality enables businesses to scale their sales operations while reducing costs dramatically.

Second, with IgniteUps.AI, businesses can replace entire sales teams with AI agents that operate at 1% of the cost of a single human employee. This makes the platform particularly valuable for startups and small businesses looking to compete with larger organizations.

Finally, IgniteUps.AI integrates with CRMs and other sales tools, ensuring that customer data is synced and accessible across all platforms.

CHOOSING THE RIGHT TOOLS FOR YOUR TEAM

When selecting AI sales tools for your team, consider the following:

Choosing the Right Tools
for Your Team

INTEGRATION

Does the tool work seamlessly with your existing systems?

EASE OF USE

Will your sales team adopt the tool, or is it too complicated?

SCALABILITY

Can the platform grow with your business?

ROI

Does the tool justify its cost through improved efficiency and revenue?

A well-rounded sales tech stack might include CRM to manage customer relationships and store data (e.g., Salesforce), an AI sales platform, to automate tasks and provide insights (e.g., IgniteUps.AI), a data hygiene tool to maintain clean, accurate data (e.g., Spearphish.io), and a communication tool to facilitate outreach and engagement (e.g., Outreach.io). By combining these tools, businesses can create a seamless, AI-powered sales operation.

THE FUTURE OF SALES PLATFORMS

The future isn't about replacing sales professionals with AI—it's about empowering them. Tools will focus on enhancing human skills, enabling reps to focus on relationship-building and problem-solving. But the right tools and platforms will be essential when it comes to leveraging AI in sales. By selecting technologies such as IgniteUps.AI and Spearphish.io, businesses will be able to streamline operations, maximize efficiency, and empower their teams to perform at their best.

> "AI is sometimes incorrectly framed as machines replacing humans. It's not about machines replacing humans but machines augmenting humans."
>
> —*Robin Bordoli, Partner at Authentic Ventures*

Next, let's look at how businesses can harness the power of these tools to create a future-proof sales strategy that balances automation with human expertise.

CHAPTER 13

The Future of Sales with AI

The sales landscape has never stood still, and with the rise of artificial intelligence, it's undergoing its most transformative shift yet. While traditional methods have been about building relationships, negotiation, and persistence, AI is fundamentally redefining the way sales professionals operate, interact, and succeed.

The future of sales isn't about replacing human professionals with AI but rather enhancing their capabilities. It's about balancing automation with the human touch, leveraging predictive insights, and creating hyper-personalized experiences.

PREDICTIVE SELLING
ANTICIPATING CUSTOMER NEEDS

Predictive selling leverages AI to analyze historical data, customer behaviors, and market trends to anticipate what customers need, even before they realize it. Instead of reacting to customer inquiries, sales teams can proactively offer solutions, readily positioning themselves as problem-solvers and trusted advisors. Platforms like Salesforce Einstein and HubSpot leverage AI to provide predictive insights, helping sales teams focus on high-value opportunities and tailor their approach for maximum impact.

For example, an AI-powered CRM might alert a salesperson that a customer is nearing the end of their subscription and recommend outreach in order to discuss renewals or upgrades. Also, AI can identify patterns that indicate when a prospect is likely to buy, such as frequent visits to certain product pages or opening multiple emails in a short period.

Predictive selling offers numerous advantages, and one of those advantages is shortened sales cycles. By identifying when customers are ready to buy, sales teams can engage at the optimal time. It also offers increased conversions, as proactive outreach builds trust and demonstrates value, increasing the likelihood of closing deals. Last but not least: enhanced customer experience. Having the ability to anticipate customer needs shows that your business understands and values your customers.

Put simply, today's customers expect personalized experiences; generic emails and cookie-cutter pitches no longer suffice. AI enables hyper-personalization by analyzing individual customer data and delivering tailored messaging, recommendations, and solutions. For example, AI can recommend specific products based on a customer's browsing history and preferences. Additionally, email campaigns can be dynamically adjusted to include a customer's name, past purchases, and even references to their unique challenges.

Traditionally, personalization was a labor-intensive process that only top-tier customers received. AI automates this process, allowing businesses to deliver personalized experiences to thousands—or even millions—of customers simultaneously.

CASE STUDY
Personalization in Action

A retail company used AI to personalize product recommendations for their e-commerce site. The results included a 25% increase in average order value, a 40% boost in repeat purchases, and higher customer satisfaction, with personalized recommendations cited as a top factor.

AI AS A CATALYST FOR
MULTI-CHANNEL SALES

In today's sales landscape, customers interact with brands across multiple channels: email, phone, social media, chat, and more. And AI allows sales teams to maintain consistency and relevance across all these touchpoints.

AI-powered chatbots can provide instant support on a website while collecting data that informs follow-up emails or calls. Further, social listening tools powered by AI can identify prospects discussing relevant topics on platforms like LinkedIn or X, allowing sales teams to engage at the right moment.

AI tools can also automate and synchronize multi-channel outreach, ensuring that prospects receive the right message at the right time, no matter where they are in their buyer journey. Platforms like IgniteUps.AI excel in automating communication across multiple channels, saving time and improving engagement.

As AI handles routine tasks like lead scoring and follow-ups, the role of the sales professional is shifting toward more of a consultative approach. Salespeople are no longer just closing deals; they're building relationships, guiding customers through complex decisions, and delivering long-term value.

But while AI can automate many processes, over-reliance on it can lead to impersonal interactions. Sales teams must strike a balance between automation and the human touch to ensure customers feel valued. To thrive in an AI-driven world, sales professionals *must* develop skills that complement technology, including emotional intelligence, strategic thinking, and tech savviness.

AI bridges the gap between sales and marketing by providing a unified view of the customer journey. Shared insights enable both teams to create cohesive campaigns, align messaging and strategies, and share data to improve targeting and personalization.

AI doesn't just help sales teams close deals. It also supports post-sale efforts by identifying upsell and cross-sell opportunities, monitoring customer satisfaction through sentiment analysis, and automating follow-ups to ensure customers remain engaged.

REAL-WORLD PREDICTIONS FOR THE FUTURE OF SALES

The shift toward remote work and digital communication has accelerated the adoption of virtual selling. AI will only continue to play a pivotal role by enhancing video calls with real-time insights and recommendations as well as providing virtual backgrounds that include key customer data for reference during meetings. Meanwhile, AI tools like Gong.io and Chorus.ai are evolving into virtual mentors, providing sales reps with instant feedback on their performance and actionable advice to improve future interactions.

As a result, sales managers will increasingly rely on AI to predict which team members need additional support or training, identify strategies that drive the best results, and allocate resources more effectively to maximize team performance.

PREPARING FOR THE AI-DRIVEN FUTURE

As already discussed, success in the AI-driven sales landscape depends on utilizing the right platforms. Businesses must evaluate their needs and invest in tools that seamlessly integrate with existing systems, provide actionable insights, and support multi-channel engagement. Sales professionals must also commit to continuous learning, staying updated on AI advancements, industry trends, and best practices. The combination of human expertise and AI-driven capabilities will be key to long-term success.

AI is not the end of sales as we know it—it's a new beginning. By automating routine tasks, providing predictive insights, and enabling hyper-personalization, AI will continue to empower sales teams to focus on what truly matters: building relationships, solving problems, and delivering exceptional value to customers.

The future of sales belongs to those who embrace the partnership between humans and machines. With the right tools, clean data, and a commitment to continuous improvement, businesses can thrive in the AI-driven era and redefine what it means to sell.

In the final section of this book, we'll discuss how to blend AI and human expertise to create a sales strategy that is future-proof, scalable, and primed for success.

PART 4

The New Sales Landscape

Artificial intelligence has brought sales into a new era of efficiency, scale, and precision. But as sales processes evolve, so do the complexities that come with managing data, maintaining compliance, and ensuring trust in an increasingly digital world. Beyond simply implementing AI, businesses now must navigate the challenges of data privacy, security, and regulatory compliance to stay ahead in this landscape.

In this section, we'll explore how platforms like Spearphish.io and IgniteUps.AI are reshaping the sales process by not only offering cutting-edge AI capabilities but also prioritizing critical compliance standards such as SOC 2 and TCPA to help business have scalable, compliant, and future-proof sales operations.

CHAPTER 14

Platforms Changing the Game

Artificial Intelligence is revolutionizing the sales industry, but its success hinges on the tools and platforms businesses choose to implement. The right platforms don't just enhance productivity, they also transform the way teams engage with customers, manage data, and drive revenue.

In this chapter, we'll spotlight two industry-leading platforms, Spearphish.io and IgniteUps.AI, which are setting the standard for AI-powered sales. These platforms not only optimize sales processes but also prioritize critical factors like data hygiene, scalability, and regulatory compliance.

THE ROLE OF PLATFORMS IN THE AI SALES REVOLUTION

AI's ability to analyze data, automate workflows, and personalize interactions is only as strong as the platform that powers it. A robust platform acts as the bridge between AI tools and business outcomes, offering seamless integration with existing systems, advanced automation to streamline repetitive tasks, scalability to adapt to growing sales needs, and security and compliance to protect sensitive customer data. Without the right platforms, businesses risk underutilizing AI or even facing costly inefficiencies and compliance risks.

> "AI enables marketers to... reduce time spent on repetitive, data-driven tasks."
>
> —*Paul Roetzer, Author of* Marketing Artificial Intelligence

Modern sales platforms must address key pain points, including:

Data Management: Ensuring CRM data is clean, accurate, and actionable.

Automation: Reducing manual workloads while maintaining a personal touch.

Compliance: Adhering to regulations like SOC 2 and TCPA to avoid legal exposure.

Scalability: Supporting growth without overwhelming resources.

Platforms like Spearphish.io and IgniteUps.AI excel in these areas, empowering sales teams to deliver exceptional results.

SPEARPHISH.IO
CLEANING AND OPTIMIZING
YOUR DATA

Dirty data—characterized by inaccuracies, duplicates, and missing information—undermines the performance of AI tools and wastes valuable resources. Poor data quality often leads to biased predictions, flawed decision-making, and increased processing time. AI models trained on unreliable data struggle with accuracy, causing misleading insights and damaging user trust.

Additionally, organizations must invest more time and money in data cleaning, slowing innovation and reducing efficiency. In sectors like healthcare and finance, incorrect data can have severe consequences, from misdiagnoses to financial losses. Ensuring high-quality data is essential for AI to deliver reliable, fair, and impactful results.

Spearphish.io is a data hygiene platform specifically designed to ensure your CRM data is clean, complete, and ready to power AI systems.

SPEARPHISH.IO

KEY FEATURES

DATA CLEANING

Identifies and removes duplicate, outdated, or inaccurate records.

DATA APPENDING

Adds missing contact details like phone numbers, email addresses, and physical addresses.

DATA VERIFICATION

Validates the accuracy of customer information to ensure reliability.

SOC 2 COMPLIANCE

Ensures secure handling of customer data, building trust and safeguarding privacy.

By integrating with your CRM, Spearphish.io provides immediate benefits such as enhanced AI performance, improved targeting, and reduced risk. In the area of enhanced performance, clean data allows AI tools to deliver accurate insights and predictions. Complete and accurate customer profiles enable more effective outreach, making targeting more effective. And its compliance features minimize legal exposure and protect your reputation.

CASE STUDY
A Retailer's Data Hygiene Success

A national retail chain used AI to clean and append its CRM data, which was riddled with inaccuracies. After implementation, email open rates increased by 40%, sales teams spent 30% less time correcting data and more time closing deals, and TCPA compliance violations dropped to zero, avoiding costly penalties.

IGNITEUPS.AI
AI-POWERED SALES AUTOMATION

Scaling traditional sales teams is both expensive and resource-intensive. Common challenges include high overhead costs for staffing and infrastructure, inconsistent performance across team members, and limited availability for customer interactions—especially outside business hours.

IgniteUps.AI addresses these issues by providing AI-powered sales agents who are able to handle both inbound and outbound communications at a fraction of the cost.

IGNITEUPS.AI

KEY FEATURES

AI-POWERED VIRTUAL AGENTS

Handle calls, emails, and texts 24/7 with consistent quality and efficiency.

COMPLIANCE

Ensures adhestion to TCPA regulations by flagging phone numbers on Do Not Call (DNC) lists.

TCPA COMPLIANCE

Automates outreach while adhering to telemarketing regulations, reducing legal risks.

COST EFFICIENCY

Operates at just 1% of the cost of traditional sales teams, making it ideal for businesses of all sizes.

IgniteUps.AI excels in both inbound and outbound sales. In the area of inbound sales, AI agents engage with customers in real time, answering questions, providing personalized recommendations, and capturing leads. And when it comes to outbound sales, the platform automates prospecting, follow-ups, and outreach campaigns while maintaining a personalized touch.

CASE STUDY
Scaling a SaaS Company with AI

A SaaS company specializing in customer support automation leveraged AI to scale operations and improve efficiency. By integrating AI-powered chatbots and virtual assistants, they reduced response times by 70% and handled 5x more customer inquiries—all without increasing staff.

AI-driven analytics helped personalize user experiences, which boosted retention by 30%. Also, predictive modeling optimized pricing strategies, which increased conversions by 25%.

With AI automating routine tasks, the company was able to reallocate resources to innovation and customer success, driving a 40% revenue growth in one year. This AI-driven approach enabled rapid, cost-effective scaling while enhancing customer satisfaction and business performance.

WHY SOC 2 AND TCPA COMPLIANCE ARE ESSENTIAL

SOC 2 certification ensures that a platform manages customer data securely and responsibly. This is critical for building trust with customers, as it protects sensitive information, prevents data breaches that could harm a business and its reputation, and meets the expectations of clients in highly regulated industries like finance and healthcare. Thankfully, both Spearphish.io and IgniteUps.AI prioritize SOC 2 compliance, giving businesses confidence in their security practices.

The Telephone Consumer Protection Act (TCPA) governs the ways businesses can contact customers via phone, email, and text. Violating TCPA regulations can result in costly fines, which can range from $500 to $1,500 per violation, as well as legal action that drains resources and damages a company's reputation.

IgniteUps.AI ensures TCPA compliance by automatically flagging DNC-listed phone numbers, managing opt-ins and opt-outs to respect customer preferences, and automating compliant outreach to minimize risk.

BUILDING A FUTURE-PROOF SALES STRATEGY

Spearphish.io and IgniteUps.AI work together seamlessly to create a robust, AI-powered sales strategy. Spearphish.io ensures that your data is clean and compliant, providing a solid foundation for AI-driven insights, and then IgniteUps.AI uses that data to automate and enhance sales interactions, driving efficiency and scalability.

By leveraging these platforms, businesses can scale their sales efforts without adding headcount, deliver personalized and compliant customer experiences at scale, and stay ahead of competitors by adopting cutting-edge technology and best practices.

Spearphish.io and IgniteUps.AI represent the future of sales technology, offering tools that are not only powerful but also secure and compliant. By investing in these platforms, businesses can optimize their operations, scale effectively, and build lasting trust with their customers.

The right platform is more than just a tool—it's a strategic partner that empowers your team to achieve extraordinary results. As we move forward in this AI-driven sales era, tools like these will be essential for businesses looking to stay competitive, compliant, and innovative.

In the next chapter, we'll explore how automation and AI can be seamlessly integrated into sales strategies to maximize both efficiency and human potential.

CHAPTER 15

Inbound and Outbound Reinvented

AI-Powered Call Centers

The traditional call center model—characterized by large teams of agents handling inbound inquiries and conducting outbound campaigns—is no longer sufficient to meet the demands of modern sales. Customers expect faster responses, personalized communication, and seamless experiences across channels. At the same time, businesses face increasing pressure to reduce costs and maintain compliance with data and telemarketing regulations.

AI-powered call centers have emerged as the solution to these challenges. By leveraging artificial intelligence, platforms like IgniteUps.AI are revolutionizing inbound and outbound communication, automating processes while delivering exceptional results at a fraction of the cost of traditional call centers.

THE LIMITATIONS OF TRADITIONAL CALL CENTERS

Traditional call centers require significant investment in staffing, training, infrastructure, and technology. And as businesses grow, scaling these operations becomes increasingly expensive.

Traditional call centers also struggle to handle sudden spikes in demand, such as during promotional campaigns or seasonal surges. Without sufficient resources, businesses risk losing leads or frustrating customers with long wait times.

Add to that the fact that human agents often deliver inconsistent results due to factors such as variability in skill levels and training, fatigue and high turnover rates, and difficulty maintaining consistent messaging across a large team.

AI-powered call centers like IgniteUps.AI, on the other hand, can handle unlimited inbound and outbound interactions simultaneously. And unlike human agents, AI doesn't require salaries, benefits, or training, thus enabling businesses to scale their operations without breaking the bank.

For example, a business that previously needed 50 agents to manage outbound campaigns can now achieve the same results, if not better, with IgniteUps.AI—at just 1% of the cost.

AI agents deliver consistent results every time. They use standardized scripts optimized for conversions; provide accurate, personalized responses based on customer data; and never tire, ensuring consistent performance across all interactions.

AI-powered call centers can also seamlessly manage communication across multiple channels, including phone calls, text messages, emails, and chatbots on websites or social media platforms. This multi-channel approach ensures that customers can engage with your business in their preferred way, enhancing the overall experience.

HOW IGNITEUPS.AI IS REDEFINING INBOUND SALES

IgniteUps.AI uses AI agents to handle inbound inquiries 24/7, ensuring that no customer is left waiting. Agents are able to instantly answer frequently asked questions, provide personalized product recommendations based on customer history, and escalate complex issues to human agents when necessary.

For example, a customer browsing a website at midnight can use the live chat function powered by IgniteUps.AI to get immediate answers about pricing, availability, or features.

Another great feature of IgniteUps.AI is that it doesn't just react to inbound inquiries. It also proactively engages customers by sending follow-up messages after purchases, recommending complementary products or services, and offering promotions tailored to individual preferences. And of course, IgniteUps.AI ensures that all inbound communications are handled securely and in compliance with SOC 2 standards. Customer data is protected at every stage of the interaction, building trust and ensuring regulatory compliance.

HOW IGNITEUPS.AI IS REVOLUTIONIZING OUTBOUND SALES

IgniteUps.AI analyzes customer data to prioritize leads based on their likelihood to convert. This ensures that AI agents focus on high-value prospects, maximizing the efficiency of outbound campaigns.

AI agents craft tailored messages for each prospect, using insights from CRM data. For example, a lead who recently visited your website might receive a personalized email referencing specific products they viewed. Or, a prospect in the final stages of their buyer journey might get a call offering an exclusive discount.

CASE STUDY

A healthcare provider leveraged an AI-powered call center to launch a TCPA-compliant outbound campaign targeting prospective patients for preventive care appointments.

AI-driven automation ensured adherence to regulations while optimizing call timing and messaging. Within three months, the campaign increased patient engagement by 45%, scheduled 30% more appointments, and reduced compliance risks by 80%. Additionally, the provider reduced operational costs by 40% and was able to reallocate staff to higher-value tasks.

By combining AI with compliance safeguards, the healthcare provider scaled outreach efficiently while maintaining trust and regulatory adherence.

Modern customers expect immediate responses to their inquiries. AI-powered call centers meet this expectation by providing instant, accurate answers across all communication channels. While AI agents handle most interactions, they seamlessly hand off complex cases to human agents when necessary, ensuring that customers always receive the level of support they need.

THE FUTURE OF CALL CENTERS AI AND HUMAN COLLABORATION

AI-powered call centers don't eliminate the need for human agents; they simply redefine their role. Human agents can focus on handling complex, high-stakes interactions; building deep relationships with key accounts; and strategizing and optimizing sales campaigns. Tools like IgniteUps.AI act as virtual assistants, providing human agents with real-time insights and recommendations during customer interactions. In addition, they automate administrative support, such as scheduling follow-ups and logging interactions, as well as offer data-driven coaching to improve performance over time.

> "There will be jobs lost, but also gained, and changed. The number of jobs gained and changed is going to be a much larger number."
>
> —*James Manyika, Senior Vice President of Research, Technology and Society at Google*

REAL-WORLD RESULTS
AI-POWERED CALL CENTERS IN ACTION

CASE STUDY 1

Scaling a National Retail Chain

A national retailer implemented AI-driven chatbots to manage inbound customer inquiries during a major promotional period. As a result, the AI system efficiently handled 70% of customer queries, leading to a 60% reduction in response times. This automation allowed human agents to focus on complex issues, which improved overall customer satisfaction by 25%. Additionally, the retailer experienced a 40% decrease in operational costs associated with customer support during the promotion.

This successful integration of AI technology enabled the retailer to maintain high-quality customer service while effectively managing increased inquiry volumes during peak promotional periods.

CASE STUDY 2
Boosting Outbound Sales for a SaaS Company

A national retailer implemented AI-driven virtual assistants to boost outbound sales for a SaaS company. The AI assistants engaged leads via email and SMS, automating initial outreach and follow-ups.

Within six months, lead engagement increased by 50%, and qualified leads rose by 35%. This automation allowed sales representatives to focus on high-value prospects, resulting in a 25% increase in conversion rates. Additionally, the company experienced a 40% reduction in the time spent on lead qualification, leading to more efficient sales cycles.

AI-powered call centers are transforming both inbound and outbound sales while delivering unprecedented efficiency, scalability, and compliance. By automating routine tasks, enhancing personalization, and ensuring adherence to regulations, these platforms empower businesses to focus on what matters most: building relationships with customers and driving growth.

Next, we'll explore how automation and personalization come together to create scalable, high-impact sales strategies that combine efficiency with the human touch.

CHAPTER 16

Automation Meets Personalization
AI's Magic Formula

In the world of sales, automation and personalization have traditionally been viewed as opposing forces. Automation promised speed and scale, while personalization focused on the one-to-one connections that build trust and loyalty. With the advent of artificial intelligence, however, businesses no longer have to choose between the two. AI has unlocked the ability to deliver hyper-personalized experiences at scale, creating a sales strategy that is both efficient and impactful.

This chapter explores how automation and personalization work together to transform sales, the role of AI in striking the perfect balance, and how certain platforms are empowering businesses to build meaningful customer relationships while driving growth.

THE MYTH OF IMPERSONAL AUTOMATION

In its early days, sales automation was synonymous with generic, one-size-fits-all approaches. Automated emails, robocalls, and impersonal follow-ups often left customers feeling like just another number in a database. This impersonal approach eroded trust, leading to lower engagement and conversion rates.

AI has redefined automation by enabling personalization at scale. Modern AI tools analyze customer data to craft tailored messaging, recommendations, and strategies for each individual prospect. The result is an experience that feels personal, even when delivered through automated systems. For example, AI can generate personalized email subject lines based on a prospect's recent interactions with your brand. Additionally, AI-driven chatbots can address specific customer concerns, referencing previous inquiries or preferences.

By combining automation with personalization, AI ensures that every touchpoint is relevant, timely, and impactful.

HOW AI ENABLES PERSONALIZATION AT SCALE

AI tools analyze vast amounts of data to identify patterns, preferences, and behaviors. This enables businesses to segment audiences based on specific characteristics, predict customer needs and recommend relevant products or services, and personalize messaging for each individual prospect.

One of AI's greatest strengths is its ability to deliver personalization in real time. For example, during a live chat, an AI-powered bot can reference a customer's recent purchases to suggest complementary products. During a phone call, an AI system can provide the sales rep with real-time insights about the prospect's industry, challenges, and goals.

Platforms like IgniteUps.AI excel at real-time personalization, ensuring that every interaction is tailored to the customer's unique context, whereas tools like HubSpot and Outreach.io use customer data to craft highly personalized email campaigns. Subject lines can include the recipient's name or reference their recent activity, email content can highlight products or services based on the recipient's preferences, and follow-up sequences can adjust based on whether the recipient opened the previous email or clicked a link.

AI can dynamically adjust website content based on the visitor's behavior. For instance, a first-time visitor might see an introductory offer, while a returning customer sees recommendations based on their past purchases. Geographic data can tailor content to reflect regional preferences or trends.

IgniteUps.AI is one platform that automates follow-ups across multiple channels, such as sending a personalized text message after a missed call, following up with a relevant offer after a customer downloads a whitepaper, and reminding prospects about expiring promotions through personalized emails. This approach ensures that no lead is left unattended, increasing the likelihood of conversion.

BALANCING AUTOMATION AND THE HUMAN TOUCH

Automation is ideal for repetitive, time-consuming tasks such as sending initial outreach emails, scheduling follow-ups and reminders, and managing lead scoring and segmentation. By automating these processes, sales teams can focus on high-value activities.

However, while AI handles routine tasks, there are moments when human interaction is essential, such as building trust and rapport during complex deal negotiations; addressing nuanced customer concerns that require empathy; and crafting creative, out-of-the-box solutions for unique challenges. AI tools ensure that human agents can step in seamlessly when needed, creating a collaborative approach.

THE FUTURE OF AUTOMATION AND PERSONALIZATION

AI is moving toward predictive personalization, where systems anticipate customer needs before they arise. Also, AI systems are constantly improving through machine learning. As they analyze more data, they become better at delivering personalization, identifying trends, and optimizing outreach strategies.

Automation and personalization are no longer at odds. AI has brought them together to create a magic formula for sales success. By leveraging specific AI platforms, businesses can automate routine tasks, deliver personalized experiences at scale, and maintain compliance with critical regulations.

The future of sales lies in balancing efficiency with connection, and AI makes this balance possible. By embracing these tools and strategies, businesses can build lasting relationships, drive growth, and lead the way in the AI-driven sales era. In the next chapter, we'll explore how these principles translate into cost savings and scalability, showing how AI can deliver extraordinary results for businesses of all sizes.

CHAPTER 17

Cost Efficiency

How AI Replaces Teams at 1% of the Cost

Traditional sales teams have long been a cornerstone of business success, but they come with significant costs. From salaries and benefits to training and infrastructure, building and maintaining a high-performing sales team can be an expensive endeavor. Add to that the inefficiencies of manual processes and inconsistent performance, and it becomes clear why many organizations struggle to scale their sales efforts without breaking the bank.

Enter artificial intelligence, where platforms are revolutionizing sales by automating processes, streamlining operations, and delivering consistent, high-quality results—all at a fraction of the cost of traditional sales teams.

In this chapter, we'll explore how AI-powered solutions can replace or supplement human teams, offering unmatched cost efficiency and scalability while maintaining compliance and quality.

> "AI is poised to augment rather than replace sales roles, empowering salespeople with tools for efficiency and insight."
>
> —*Kyle Coleman, CMO at Copy.ai*

THE HIGH COSTS OF TRADITIONAL SALES TEAMS

Sales teams require substantial financial investment. Competitive pay and benefits are necessary to attract and retain talent. New hires often require weeks or months of training before they can perform at a high level. Offices, equipment, and software licenses only add to the operational expenses. A mid-sized sales team of twenty people can easily cost a company over $2 million annually when factoring in all these expenses. In addition to direct costs, traditional teams often face inefficiencies that drain resources, such as high turnover rates, leading to constant recruitment and retraining; time wasted on administrative tasks like data entry and lead tracking; and inconsistent performance, with top performers often carrying the weight of the team.

As businesses grow, scaling a traditional sales team requires significant investment in additional headcount and infrastructure. This approach is not only costly but also time-intensive, making it difficult for businesses to respond quickly to market demands.

HOW AI DELIVERS COST EFFICIENCY

AI excels at automating both repetitive and time-consuming tasks, such as lead scoring and prioritization, scheduling follow-ups and reminders, and logging and tracking customer interactions. By handling these tasks, AI reduces the need for administrative staff and allows sales reps to focus on high-value activities like closing deals.

Unlike human teams, AI-powered solutions can scale effortlessly. Whether managing 100 leads or 10,000, top-notch AI platforms handle the workload without requiring additional resources or infrastructure. A traditional call center handling 10,000 monthly calls might require 50 agents. An AI-powered system, on the other hand, can handle the same volume—or more—at a fraction of the cost, with no need for additional hires.

AI also eliminates the variability inherent in human teams. AI agents deliver consistent, high-quality interactions every time, ensuring that customers receive the same level of service regardless of volume or complexity.

IGNITEUPS.AI
THE 1% SOLUTION

IgniteUps.AI provides virtual sales agents that handle inbound and outbound calls, emails, and texts with remarkable efficiency. These AI agents operate 24/7, ensuring no lead or customer is left unattended. They also deliver personalized interactions based on customer data and behavior and handle tasks at just 1% of the cost of a traditional human sales team.

A business that previously spent $500,000 annually on a traditional sales team can implement IgniteUps.AI for as little as $5,000 per year, achieving similar—or better—results. These savings free up resources for other strategic initiatives, such as marketing or product development.

IgniteUps.AI integrates with existing CRMs and communication tools, allowing businesses to deploy AI agents quickly and without disruption. This ensures a smooth transition from traditional to AI-driven sales operations.

CASE STUDIES
Real-World Success with AI

A national retail chain replaced its outbound call center with an AI platform. The results included a 90% reduction in operational costs, a 40% increase in outbound response rates (driven by personalized AI outreach), improved customer satisfaction (as AI agents ensured consistent follow-ups), and timely responses.

A SaaS company struggling to scale its sales team implemented an AI platform in order to handle inbound inquiries. Within just six months, costs dropped by 85%, lead response times improved by 50%, and the company expanded its customer base without hiring additional staff.

THE FUTURE OF COST EFFICIENCY IN SALES

AI isn't just a cost-saving tool—it's a strategic investment that drives growth and scalability. Businesses that adopt AI-powered solutions gain a competitive edge by reducing operational costs, improving sales efficiency and effectiveness, and freeing up resources to focus on innovation and customer experience.

While AI handles routine tasks, human sales professionals remain essential for building relationships with key accounts, navigating complex negotiations, and providing creative solutions to unique challenges.

As I've already stated, the future of sales is a collaborative model where AI and humans work together to achieve extraordinary results, and AI platforms are redefining what's possible in sales by delivering exceptional results at just 1% of the cost of traditional teams. By automating routine tasks, scaling effortlessly, and maintaining compliance, AI-powered solutions empower businesses to achieve more with less.

The economics of AI-driven sales are clear: lower costs, higher efficiency, and better outcomes. For businesses looking to stay competitive in a rapidly evolving market, adopting AI is no longer optional—it's essential.

BEST PRACTICES FOR ENSURING COMPLIANCE

Educate your sales team on compliance requirements, such as gaining consent, respecting opt-outs, and protecting customer data. A well-informed team reduces the risk of accidental violations. Also, regular audits of your sales processes and data ensure ongoing compliance. Be sure to identify and address issues before they become liabilities. Compliance isn't just about avoiding fines—it's about building trust, protecting your brand, and ensuring long-term success.

As AI continues to evolve, compliance tools will become even more sophisticated. Predictive algorithms will identify potential risks before they occur, and real-time monitoring will ensure that every interaction meets regulatory standards.

With increasing globalization, most businesses have to prepare to navigate multiple regulatory environments, such as GDPR in Europe or CCPA in California. Thankfully, platforms like Spearphish.io and IgniteUps.AI are already equipped to help businesses manage these complexities.

By investing in the right tools, properly training your team, and prioritizing compliance, you can create a sales operation that's not only efficient but also trustworthy and future-proof. The new sales landscape demands responsibility as much as it does innovation, and those who embrace compliance will lead the way into the future.

In the final part of this book, we'll explore how businesses can combine everything we've learned—AI, compliance, personalization, and scalability—to build a winning sales strategy for the AI-driven era.

PART 5

Capitalizing on AI Sales Platforms

AI is no longer a "nice-to-have" for sales teams—it's a necessity for those looking to thrive in an increasingly competitive and fast-paced marketplace.

AI has revolutionized the way businesses interact with customers, manage data, and close deals. However, success in the AI-driven sales era requires more than just adopting the right tools. It demands a clear strategy, a focus on compliance, and a commitment to leveraging AI to enhance, not replace, the human element of sales.

We will now bring together all the insights and strategies discussed throughout the book. We'll explore how to maximize the potential of AI-powered platforms, why clean data and compliance remain critical, and how businesses can blend AI with human expertise to create a winning formula for sales success.

CHAPTER 18

Mastering AI Platforms
Spearphish.io

In the rapidly evolving world of sales, artificial intelligence platforms are only as effective as the data they process. Remember, AI thrives on clean, accurate, and complete data to deliver the insights, automation, and personalization that drive sales success. Yet many organizations overlook the importance of data hygiene, relying on outdated, incomplete, or inaccurate customer information. This not only hampers AI performance but also limits the effectiveness of sales strategies.

This is where artificial intelligence platforms such as Spearphish.io come in. By ensuring data is clean, compliant, and ready to use, Spearphish.io has become an indispensable tool for businesses aiming to maximize the potential of AI in sales.

In this chapter, we'll explore how to fully

leverage Spearphish.io to transform your CRM into a powerful sales engine.

WHY SPEARPHISH.IO IS A GAME-CHANGER

Studies show that 30% of CRM data becomes outdated each year due to changes in customer contact information, job roles, or preferences. This "dirty data" leads to wasted resources as sales teams spend time pursuing dead leads or correcting errors instead of closing deals. It also leads to missed opportunities, given that incomplete profiles make it difficult to deliver personalized experiences. Also, compliance risks. After all, outdated contact information can lead to violations of regulations.

Spearphish.io ensures that your CRM is always clean, accurate, and compliant. Its key features include data cleaning, data appending, data verification, and compliance support. By integrating with your CRM, Spearphish.io ensures that your sales team always has access to reliable, actionable data.

HOW TO GET THE MOST OUT OF SPEARPHISH.IO

The first step to mastering Spearphish.io is

integrating it with your existing CRM. This allows the platform to automatically clean and update customer records, append missing information to create complete profiles, and synchronize data across all systems for consistency.

Second, you want to make data hygiene a continuous process by scheduling regular audits. Spearphish.io can automatically identify and address issues such as duplicates caused by manual data entry errors, outdated phone numbers or email addresses, and incomplete records that hinder AI-driven personalization. Regular audits ensure that your data remains accurate and ready for AI processing.

Spearphish.io doesn't just clean data—it also provides actionable insights. You can easily use these insights to segment your audience based on updated demographics or behaviors, prioritize high-value leads with complete and verified contact details, and improve personalization by identifying trends and preferences in customer data.

CASE STUDY
A SaaS Company's Data Transformation

A SaaS company struggled with low email open rates and poor lead engagement due to incomplete and inaccurate CRM data. However, after integrating Spearphish.io, the platform appended missing email addresses and verified existing ones, email open rates increased by 40%, lead-to-opportunity conversion rates improved by 25%, and AI-driven outreach campaigns became more effective, delivering a 30% higher ROI.

Additionally, by automating data cleaning and compliance processes, Spearphish.io eliminates the need for manual data management. This reduces operational costs and allows sales teams to focus on revenue-generating activities.

Spearphish.io ensures that your CRM can handle growth without sacrificing data quality. Whether you're managing 1,000 or one million customer records, the platform ensures your data remains reliable and actionable.

BEST PRACTICES FOR USING SPEARPHISH.IO

Integrate Early: Don't wait until your CRM is overwhelmed with dirty data—integrate AI as early as possible to maintain clean records.

Schedule Regular Audits: Use the AI platform to conduct monthly or quarterly data hygiene checks.

Combine with AI Tools: Pair your data cleaning platform with another AI platform to maximize the impact of clean data on your sales strategy.

Educate Your Team: Train your sales team on the importance of data hygiene and how to use your chosen AI software programs effectively.

Spearphish.io isn't just a platform; it's the very foundation of a future-proof sales strategy. And when combined with AI tools like IgniteUps.AI, it transforms the way businesses engage with customers, driving growth and long-term success.

In the next chapter, we'll take the next step toward building the ultimate AI-powered sales operation by exploring how to train and deploy AI agents using IgniteUps.AI, ensuring that your sales team can capitalize on automation, personalization, and scalability.

CHAPTER 19

Training Your AI Agents with IgniteUps.AI

In an era where speed, scalability, and personalization are critical to sales success, AI-powered agents have emerged as a transformative solution. Platforms like IgniteUps.AI allow businesses to automate inbound and outbound sales processes while delivering consistent, high-quality customer interactions. But like any tool, AI agents require careful setup and optimization to achieve their full potential.

This chapter focuses on how to train and deploy AI agents using IgniteUps.AI to maximize efficiency, compliance, and customer satisfaction. From defining objectives to monitoring performance, we'll guide you through the steps to turn your AI agents into valuable assets for your sales team.

WHAT MAKES IGNITEUPS.AI A GAME-CHANGER?

AI Agents are built for sales. IgniteUps.AI provides virtual agents that can handle calls, emails, and texts 24/7; deliver personalized interactions using customer data; and respond to inquiries instantly, ensuring no lead is left waiting. And unlike human agents, these AI-powered counterparts operate with unwavering consistency and at a fraction of the cost.

KEY FEATURES OF IGNITEUPS.AI

Inbound and Outbound Automation: AI agents can manage inbound inquiries while simultaneously conducting outbound campaigns, ensuring seamless customer engagement.

Personalization at Scale: Using CRM data, IgniteUps.AI customizes interactions to individual customers' preferences and needs.

Compliance Built In: With SOC 2 and TCPA compliance, IgniteUps.AI ensures that all customer interactions meet legal and ethical standards.

PREPARING FOR AI AGENT DEPLOYMENT

In preparation for AI deployment, the first thing you must do is clean your data. (I'll say it again and again; it's that important. AI agents are only as good as the data they work with.) Before deploying IgniteUps.AI, ensure that your CRM data is accurate, complete, and up to date. Remove duplicates and outdated records, append missing contact details such as phone numbers and emails, and verify customer preferences to ensure personalized and compliant communication.

From there, determine what you want your AI agents to achieve. Common objectives include identifying high-value prospects and routing them to human agents, managing calendars/appointments and reducing friction in the sales process, supporting customers by answering FAQs and resolving basic issues, and nurturing leads through the funnel, closing deals where possible. By setting specific goals, you can tailor your AI agents to align with your business needs.

AI agents need clear, effective communication guidelines. Work with your sales team to develop scripts for common customer inquiries, responses to objections or concerns, and conversation flows that guide prospects toward the next step in the sales process. Keep scripts concise, customer-focused, and aligned with your brand voice.

TRAINING YOUR AI AGENTS

Because AI agents rely on data to make decisions and personalize interactions, ensure that your CRM provides context for each customer interaction, including purchase history, browsing behavior, and communication preferences. This allows AI agents to tailor their approach to each individual customer.

IgniteUps.AI leverages machine learning to continuously refine its performance. Machine learning ensures that your AI agents get smarter and more effective with each interaction. Here's how you can best support this process:

Monitor Interactions: Review transcripts and outcomes to identify areas for improvement.

Provide Feedback: Flag responses that need adjustment and update scripts accordingly.

Analyze Patterns: Use data insights to refine workflows and enhance agent performance over time.

Before rolling out AI agents across your entire sales operation, conduct pilot tests with a smaller group of leads. After that, gather feedback from customers and your sales team. Finally, adjust workflows and scripts based on test results. This kind of phased approach minimizes risks and ensures a smooth transition.

Once you have your system in place, evaluate the performance of your AI agents by tracking key metrics, such as response time (how quickly AI agents respond to customer inquiries), the percentage of interactions that result in a lead advancing through the sales funnel (conversion rates), feedback and reviews from customers about their interactions with AI agents, and TCPA compliance rates and error-free outreach campaigns. By regularly reviewing these metrics, you can identify strengths and areas for improvement.

AI agents are most effective when paired with skilled human sales professionals. For example, AI agents can qualify leads and pass high-priority prospects to human agents for follow-up. Then, those human agents can review AI performance data and provide insights to refine workflows. This level of collaboration ensures that your sales operation benefits from both AI efficiency and human expertise.

REAL-WORLD EXAMPLES OF AI AGENT SUCCESS

Automating Inbound Sales for an

E-Commerce Brand

An e-commerce company decided to implement AI software to handle inbound customer inquiries. The AI agents answered over 10,000 inquiries per month (with a 95% accuracy rate), resolved common questions instantly (reducing support ticket volumes by 40%), and passed complex issues to human agents (ensuring seamless escalation).

The company saved $200,000 annually in staffing costs while improving customer satisfaction scores by 30%.

Scaling Outbound Campaigns for a B2B SaaS Provider

A SaaS provider used AI software to automate outbound sales campaigns targeting small businesses. The results included a 50% increase in response rates due to personalized outreach, a 75% reduction in operational costs compared to using human agents, and zero TCPA violations, thanks to built-in compliance features.

The provider scaled its sales operations without adding headcount, achieving a 35% boost in revenue.

Best Practices

for Long-Term Success

Combine AI and Human Expertise

Use AI agents to handle routine tasks and free up human agents for high-value interactions.

Regularly Update Scripts

Refine conversation flows based on customer feedback and evolving business goals.

Leverage Data Insights

Use performance metrics provided by IgniteUps.AI to continuously optimize your sales strategy.

Scale Strategically

Expand AI deployment gradually, ensuring your infrastructure and processes can support growth.

IgniteUps.AI isn't just a tool—it's a partner that transforms the way businesses approach sales. By automating routine tasks, delivering personalized interactions, and ensuring compliance, IgniteUps.AI empowers sales teams to focus on building relationships, closing deals, and driving growth.

When paired with clean, accurate data IgniteUps.AI becomes even more powerful, enabling businesses to scale their operations efficiently and effectively. By training your AI agents and leveraging their capabilities, you can create a sales strategy that is smarter, faster, and more impactful than ever before.

Next, we'll explore how to combine these AI tools with human expertise to create a sales operation that balances automation, personalization, and relationship-building. Let's take the final step toward mastering the AI-driven sales landscape.

CHAPTER 20

The Skills You Need to Sell More with AI

The introduction of artificial intelligence into the sales industry has created a new set of opportunities—and challenges. While AI tools handle data-driven tasks, automate processes, and personalize interactions at scale, they can't replace the human qualities that make a great salesperson truly exceptional.

In this chapter, we'll explore the skills sales professionals must develop to thrive in an AI-driven world. Combining human expertise with AI capabilities is the key to building trust, creating value, and achieving extraordinary results.

EMOTIONAL INTELLIGENCE
THE FOUNDATION OF
HUMAN-CENTRIC SALES

AI excels at analyzing data and predicting outcomes, but it cannot replicate human empathy and emotional intelligence. These qualities are critical for building trust and rapport with customers, navigating complex negotiations, and understanding subtle emotional cues like hesitation or enthusiasm.

Sales professionals can enhance their emotional intelligence by focusing on skills such as:

Active Listening: Pay close attention to what the customer is saying (and not saying). Reflect on their words to ensure they feel heard.

Empathy: Put yourself in the customer's shoes to understand their challenges and emotions.

Adaptability: Adjust your tone, approach, and strategy based on the customer's mood and preferences.

Consider this example: During a negotiation, a customer may express concerns about budget constraints. While AI might suggest offering a discount, an emotionally intelligent salesperson might dig deeper to uncover the root of the concern—perhaps the customer needs assurance about long-term ROI.

STRATEGIC THINKING
TURNING AI INSIGHTS
INTO ACTION

AI provides sales teams with valuable data-driven insights, such as which leads are most likely to convert, which products or services align with a customer's needs, and the best times to reach out to prospects. However, interpreting these insights and turning them into actionable strategies requires strategic thinking.

In order to develop strategic thinking skills, salespeople must:

Understand the Big Picture: Go beyond individual deals and consider how each interaction fits into the customer's overall journey.

Prioritize High-Impact Activities: Use AI insights to focus on the leads, accounts, and tasks that offer the highest ROI.

Anticipate Challenges: Use data trends to predict potential obstacles and proactively address them.

Imagine that an AI platform identifies that a prospect is actively researching competitors. A strategic salesperson might use this information to craft a pitch highlighting their product's unique advantages, positioning their solution as the superior choice.

Tech-savviness is also essential when it comes to mastering AI tools. After all, the tools are only as effective as the people using them. To maximize the value of AI platforms such as IgniteUps.AI and Spearphish.io, sales professionals must understand how to navigate the platforms efficiently, interpret the data and insights they provide, and customize workflows to align with business goals.

BEST PRACTICES FOR BECOMING A TECH-SAVVY SALES PROFESSIONAL

Invest in Training: Take advantage of onboarding sessions, tutorials, and training resources provided by your AI platform vendors.

Stay Curious: Explore new features and updates regularly to stay ahead of the curve.

Collaborate with Tech Teams: Work closely with IT and data teams to understand how AI tools integrate with your CRM and other systems.

CASE STUDY

AI-Powered Success

A real estate company implemented a custom AI-powered tool to streamline its lead generation process, making it more efficient and cost-effective. The AI platform automated initial outreach and follow-ups, ensuring timely and personalized communication with potential clients.

Within six months, the firm experienced a 35% increase in lead engagement and a 25% boost in conversion rates. Additionally, the automation reduced the workload on sales representatives, allowing them to focus on high-value prospects, which enhanced overall productivity.

This integration of AI technology significantly improved the firm's lead nurturing process, leading to increased revenue and business growth.

ADAPTABILITY
THRIVING IN A CHANGING LANDSCAPE

AI is changing the way sales teams operate, shifting the focus from routine tasks to high-value activities. Sales professionals must be adaptable, embracing new tools and methods while maintaining the core principles of selling.

There are three key steps in cultivating adaptability. First, *learn to embrace change*. View AI as an opportunity, not a threat, and be open to learning new technologies and techniques. Second, *stay agile*. Be prepared to adjust your approach based on AI insights, customer feedback, and market trends. And finally, *experiment and innovate*. Use AI tools to test new strategies and refine your sales process.

Consider a seasoned sales rep accustomed to traditional methods who embraced AI-powered lead scoring to prioritize their pipeline. By adapting their workflow to incorporate AI recommendations, they closed 20% more deals in a quarter. The most successful sales teams treat AI as a partner rather than a replacement.

AI tools also enable better collaboration among team members by providing shared insights that align sales and marketing efforts, highlighting best practices based on performance data, and allowing managers to track team progress and identify areas for improvement.

CASE STUDY
A Collaborative Approach

AI platforms can streamline routine follow-up tasks, allowing human representatives to concentrate on more strategic activities, thereby enhancing overall productivity and potentially boosting sales.

AI-driven assistants engage leads via email, chat, and SMS, allowing human sales representatives to focus on high-value accounts. This approach has led to increased efficiency and productivity in sales teams.

There are also AI platforms that provide solutions offering real-time transcriptions and alerts during calls, helping agents manage routine follow-ups more effectively. This technology enhances agent performance and contributes to improved customer interactions.

The sales industry is constantly evolving, and staying competitive requires a commitment to continuous learning. AI tools are updated regularly, and new techniques and technologies emerge every year.

Keep learning by investing in education (attend workshops, webinars, and courses on AI-driven sales techniques); follow industry trends, staying informed about the latest advancements in sales technology and customer behavior; and seeking feedback, regularly reviewing your performance and asking for input from peers, managers, and even AI tools like Gong.io.

A sales professional who combines AI proficiency with a willingness to learn new skills will always stand out. They'll adapt faster, close more deals, and build stronger relationships with customers.

Yes, AI is reshaping the sales landscape, but the human element will continue to remain irreplaceable. By developing emotional intelligence, strategic thinking, tech-savviness, adaptability, and a commitment to continuous learning, sales professionals can thrive in this new era. The combination of human expertise and AI capabilities creates a powerful synergy that drives results, builds trust, and enhances customer experiences. As businesses continue to embrace platforms like IgniteUps.AI and Spearphish.io, the demand for skilled, AI-enhanced sales professionals will only grow.

CHAPTER 21

From Implementation to Profit

AI as a Sales Partner

Implementing artificial intelligence into your sales process is a significant milestone, but it's only the beginning. To fully capitalize on AI's potential, businesses must move beyond simple adoption and focus on integration, optimization, and execution. The ultimate goal is not just to implement AI tools like IgniteUps.AI and Spearphish.io but to turn these technologies into a profit-driving force.

In this chapter, we'll explore how to integrate AI seamlessly into your sales operations, measure success, and continuously refine your approach to maximize ROI. By treating AI as a partner rather than just another tool, you can create a high-performing, scalable sales operation that delivers measurable results.

Before implementing AI, consider what specific challenges are you trying to address (e.g., lead qualification, follow-ups, personalization); how AI will improve efficiency, customer experience, or revenue generation; and what metrics will you use to measure success. For example, if your goal is to increase lead conversion rates by 25%, you could track the percentage of leads progressing through the sales funnel before and after AI implementation.

Next, it's important to create a step-by-step plan for integrating AI into your sales process. In order to do this effectively:

Assess Your Current Tools: Ensure your CRM and existing systems can integrate with AI platforms like IgniteUps.AI and Spearphish.io.

Clean Your Data: Use Spearphish.io to prepare your data for AI by removing duplicates, appending missing details, and ensuring compliance.

Start Small: Begin with a pilot program to test AI capabilities on a smaller scale before full deployment.

Train Your Team: Provide hands-on training to ensure your sales team understands how to use the new tools effectively.

In order to evaluate the impact of AI on your sales operation, monitor the following key performance indicators (KPIs):

Lead Conversion Rates: Measure how many leads progress through the funnel and convert into customers.

Response Times: Track how quickly AI agents respond to customer inquiries.

Customer Satisfaction Scores (**CSAT**): Use surveys and feedback to assess customer experience.

Cost Per Lead: Compare the cost of acquiring leads before and after AI implementation.

ROI: Calculate the overall return on investment for AI platforms by comparing revenue gains to implementation costs.

Benchmarks provide a clear reference point for measuring progress and identifying areas for improvement. This is why it's so important to establish baseline metrics before implementing AI. For example, if your current average response time is twelve hours, perhaps your goal after AI implementation is one hour or less.

OPTIMIZING AI TOOLS FOR MAXIMUM IMPACT

AI tools are most effective when regularly monitored and optimized. Steps you can take to ensure continuous improvement include:

Review Performance Data: Analyze metrics like response rates and customer interactions to identify trends and opportunities.

Refine Scripts: Update conversation flows and messaging to align with customer feedback and evolving business goals.

Test New Strategies: Use AI to A/B test different approaches, such as varying email subject lines or call-to-action phrasing.

AI platforms leverage machine learning in order to improve over time. You can support this process by providing feedback on successful and unsuccessful interactions, allowing the AI to learn from real-world data and regularly updating algorithms to align with business objectives.

Be sure to provide ongoing training to help your team understand how to interpret AI-generated insights, use AI tools like IgniteUps.AI to streamline workflows, and focus on high-value activities that AI cannot perform, such as emotional connection and strategic decision-making.

REAL-WORLD SUCCESS STORIES
E-Commerce Success with AI

These examples highlight the tangible benefits of combining AI platforms with a clear strategy.

An e-commerce brand used AI to automate inbound customer inquiries during a major holiday campaign. The results: reduced response times from 10 hours to under a minute, increased sales by 30% through personalized product recommendations, and a savings of $150,000 in staffing costs during the campaign.

B2B SaaS Growth Through Data Hygiene

A SaaS provider integrated AI to clean and append its CRM data. The results: improved email open rates by 45%, increased lead conversion rates by 20%, and the achievement of full TCPA compliance, avoiding potential fines and legal risks.

THE FINANCIAL IMPACT OF AI

AI eliminates many of the overhead costs associated with traditional sales teams, such as salaries and benefits for additional staff, training and onboarding expenses, and manual data management efforts. Additionally, with faster response times, personalized interactions, and improved lead conversion rates, AI directly contributes to increased sales and revenue. By integrating AI into your sales strategy, measuring its impact, and continuously optimizing its performance, you can transform your sales operation into a profit-generating powerhouse.

With the insights and strategies outlined in this book, you are now equipped to capitalize on the rise of AI in sales. But the journey doesn't end here. Continue learning, experimenting, and innovating to build a sales operation that thrives in the age of AI. The future is yours to create.

PART 6

The Future of Sales in an AI-Driven World

The rise of artificial intelligence has not only transformed the sales industry, but it has also created an entirely new paradigm. Sales is no longer about who can make the most calls or deliver the slickest pitch. Instead, it's about building trust, personalizing interactions, and delivering solutions at the right moment, all powered by data-driven insights.

As we look to the future, the convergence of AI and human expertise will redefine success in sales. In this final section, we'll explore how sales teams can adapt to a rapidly changing landscape, the trends shaping the next decade, and the tools and strategies businesses need to stay competitive. The future of sales is here, so let's map out how to thrive in it.

CHAPTER 22

The Next Decade of Sales

Predictions and Trends

The sales profession has always been dynamic, adapting to shifts in technology, consumer behavior, and market demands. But with the rapid integration of artificial intelligence into nearly every aspect of sales, we're standing on the brink of unprecedented change. Over the next decade, AI will reshape the way businesses connect with customers, manage data, and drive revenue.

In this chapter, we'll explore key trends that will define the future of sales, predictions for how AI will evolve, and how sales professionals can stay ahead in this transformative era. The future isn't just about automation—it's about leveraging technology to enhance the human connection at scale.

HYPER-PERSONALIZATION BECOMES THE STANDARD

Hyper-personalization goes beyond simply addressing customers by name. It uses AI to analyze data from multiple sources—purchase history, browsing behavior, demographics, and even social media activity—to deliver highly tailored experiences at every touchpoint.

AI enables businesses to create dynamic customer profiles that update in real time, recommend products and services based on predicted needs, and tailor messaging to align with each customer's preferences, tone, and style. As an example, an AI-driven CRM could identify a customer's interest in eco-friendly products and automatically adjust marketing messages to emphasize sustainability.

Hyper-personalization builds trust and loyalty by making customers feel understood and valued, and companies that embrace this trend will stand out in an increasingly crowded marketplace.

PREDICTIVE SELLING ANTICIPATING CUSTOMER NEEDS

Traditional sales strategies are often reactive. They respond to customer inquiries and pain points. Predictive selling flips this model, enabling sales teams to anticipate customer needs before they arise.

AI uses machine learning to analyze historical data, market trends, and customer behavior, predicting which leads are most likely to convert, when customers are ready to make a purchase, and which products or services are most relevant to their needs. For instance, a SaaS company might use AI to identify customers nearing the end of their contract and proactively offer renewal incentives before competitors can reach them.

Data can drive proactive sales in myriad ways. Platforms like Spearphish.io clean and append CRM data, allowing AI tools to identify at-risk customers who may need additional attention, predict when a lead is ready to convert based on behavior patterns, and generate actionable insights for upselling and cross-selling opportunities.

CASE STUDY
Proactive Selling in Action

A financial services company used predictive insights from AI to identify customers nearing the end of their contracts. By proactively reaching out with renewal offers, they increased retention rates by 40%.

There are a number of clear benefits of predictive selling, including:

Shortened Sales Cycles: By engaging customers at the right moment, deals close faster.

Higher Conversion Rates: Anticipating needs allows for tailored pitches that resonate with customers.

Increased Customer Satisfaction: Proactive support demonstrates attentiveness and builds trust.

AI is already transforming repetitive tasks like lead scoring and follow-ups, but the next frontier of innovation that IgniteUps.AI is focused on is dynamic content creation. IgniteUps.Ai is leveraging advanced integrations into ChatGPT and Google's Gemini, and it's capable of generating personalized emails, proposals, and sales scripts tailored to each customer. Additionally, AI is monitoring live interactions, enabling real-time adjustments to tone, messaging, or strategy to optimize outcomes.

End-to-end automation will become the standard, where AI will streamline entire sales workflows—from prospecting to closing—with minimal human intervention.

However, human oversight will remain essential. While automation drives efficiency, human involvement ensures that interactions stay authentic and aligned with brand values. Sales professionals will continue to play a crucial role in refining AI strategies and handling complex or high-stakes situations that require a personal touch.

THE RISE OF
AI-POWERED COACHING

AI is also transforming sales training, and it does so by providing real-time coaching for reps during live interactions. Tools like Gong.io and Chorus.ai can analyze conversations and offer suggestions to improve tone and delivery, handle objections more effectively, as well as highlight opportunities for upselling or cross-selling. For instance, an AI tool might alert a salesperson to pause more frequently during a call or rephrase a response to better address the customer's concerns.

AI-driven coaching platforms also create personalized training programs based on each rep's strengths and weaknesses, ensuring continuous improvement.

THE RISE OF MULTI-CHANNEL
ENGAGEMENT

Multi-channel engagement will take center stage, allowing salespeople to meet customers where they are. The primary benefits of multi-channel engagement include flexibility, as customers can choose how they want to interact with your brand; personalization, with AI tailoring its messaging to suit the tone and format of each channel; and higher conversion rates, as customers are engaged where they feel most comfortable.

Customers now engage with brands across multiple channels—email, phone, chat, social media, and more. AI enables sales teams to provide consistent, seamless experiences across all these touchpoints. Within this ecosystem, AI-powered platforms like IgniteUps.AI synchronize customer interactions across channels and provide context for each interaction, ensuring continuity and personalized messaging to suit the channel as well as the customer's preferences.

As this sector continues to develop, of growing importance will be compliance and security as regulatory pressures increase. As AI adoption grows, so will regulatory scrutiny. Businesses must ensure compliance with data privacy laws (regulations such as GDPR and CCPA require strict data handling practices) and telecommunication laws. TCPA compliance will remain critical for outbound campaigns.

Thankfully, platforms like Spearphish.io and IgniteUps.AI make compliance easy by managing Do Not Call (DNC) lists automatically, verifying customer consent for outreach, and securing data with SOC 2-certified protocols.

SUSTAINABILITY AS A
SALES PRIORITY

Customers are increasingly prioritizing sustainability and ethical business practices. AI can help businesses align with these values by analyzing supply chain data to identify sustainable sourcing options, personalizing messaging to highlight eco-friendly products or practices, and reducing waste in sales processes through smarter targeting and automation.

In short, sustainability matters in sales. Aligning with customer values builds trust and loyalty, giving businesses a competitive edge in a market that rewards social responsibility.

THE EVOLVING ROLE OF THE
SALES PROFESSIONAL

As AI takes over routine tasks, sales professionals will transition to more strategic roles. They'll act as trusted advisors, providing guidance and expertise beyond product features. As relationship builders, they'll focus on long-term customer loyalty. And as creative problem-solvers, they'll craft innovative solutions to complex challenges. To thrive in the AI era, sales professionals *must* focus on emotional intelligence and empathy, strategic thinking and adaptability, and mastering AI tools and leveraging data insights effectively.

By staying ahead of trends like hyper-personalization, predictive selling, and multi-channel engagement, sales teams can position themselves as leaders in the new sales landscape. The future belongs to those who adapt, innovate, and capitalize on the opportunities that AI presents.

CHAPTER 23

Navigating the Shift
Staying Ahead in the AI Era

Artificial intelligence has fundamentally changed the rules of sales. Traditional methods of cold calling and manual lead tracking have given way to hyper-personalized, data-driven strategies powered by AI tools. While the shift to AI offers tremendous opportunities, it also presents challenges that require adaptation, innovation, and a clear focus on staying ahead of the curve.

Let's explore the practical steps businesses and sales professionals must take to thrive in the AI-driven era. Consider this your roadmap to navigating the evolving sales landscape.

CASE STUDY
Innovation in Action

A retail company used AI to test multiple email

subject lines during a holiday campaign. By identifying the best-performing option in real time, they increased open rates by 25% and boosted revenue by 15%.

As AI evolves, staying ahead requires a commitment to learning and adaptation. Keep an eye on trends such as Voice AI (tools that enable voice-activated interactions with customers), predictive analytics (advanced AI models that forecast customer behavior with greater accuracy), and dynamic content creation (AI-generated content tailored to individual customers in real time).

Platforms like IgniteUps.AI and Spearphish.io are constantly updating their features to stay ahead of the curve. Partnering with these innovators ensures your business remains competitive.

Also, be sure to join industry forums, attend conferences, and collaborate with other professionals to stay informed about the latest advancements and best practices in AI-driven sales.

Navigating the shift to AI-driven sales requires more than just adopting new tools—it demands a mindset of continuous learning, innovation, and adaptability. Provided they embrace multi-channel engagement, leverage data for proactive strategies, and build future-ready teams, businesses can stay ahead in the AI era.

CHAPTER 24

Human Creativity Meets AI Precision
The Perfect Blend

As artificial intelligence continues to revolutionize the sales industry, one truth has become clear: AI is not here to replace humans; it's here to enhance what humans do best. While AI excels at analyzing data, automating repetitive tasks, and delivering real-time insights, it lacks the critical qualities only humans can provide. Together, human creativity and AI precision form a partnership that drives extraordinary results.

This chapter explores how businesses can integrate human expertise with AI-driven capabilities to create a sales process that is efficient, personalized, and impactful. By leveraging the strengths of both, organizations can deliver exceptional customer experiences and achieve unprecedented growth.

WHY HUMANS AND AI ARE BETTER TOGETHER

AI offers advantages that humans cannot replicate. For one, its speed and efficiency. AI can analyze vast amounts of data in seconds, identifying patterns and trends that would take humans weeks to uncover. It's also highly scalable. AI tools can handle thousands of interactions simultaneously, making it easier to manage large volumes of leads. And let's not forget consistency. Unlike humans, AI delivers consistent performance, ensuring uniform quality across all interactions.

That being said, humans bring qualities to the table that AI cannot replicate. Sales professionals excel at reading emotions, building trust, and adapting to subtle cues. Humans can think outside the box, crafting innovative solutions to complex problems. And long-term customer loyalty is built on genuine human connections.

When combined, humans and AI create a powerful partnership. AI handles repetitive tasks and provides insights, freeing up humans to focus on creativity and relationship-building, and humans use their judgment and empathy to interpret AI insights and deliver personalized, meaningful interactions.

AI automates the mundane, taking care of repetitive, time-consuming tasks such as lead scoring and prioritization, scheduling follow-ups and reminders, and logging and analyzing customer interactions. Some platforms automate outbound campaigns, ensuring that every lead receives timely, personalized communication without requiring manual effort while others provide actionable insights by cleaning and analyzing CRM data. Sales professionals can use these insights to focus on high-priority leads, tailor their approach based on customer preferences, and identify opportunities for upselling or cross-selling.

CASE STUDY
Efficiency in Action

A financial technology firm integrated an AI sales simulator and coaching software to enhance their sales process. This AI-driven platform provided realistic simulations for sales representatives, allowing them to practice and refine their skills in a controlled environment.

As a result, the firm achieved a 13.97% improvement in sales cycle progression. The platform enabled comprehensive training sessions, with sales reps completing an average of 11 practice attempts each, surpassing traditional role-playing methods. This rigorous training approach led to more effective sales conversations, contributing to increased efficiency and productivity within the sales team.

CREATIVITY IN PERSONALIZATION

AI-powered platforms analyze customer data to create detailed profiles, enabling personalized email campaigns tailored to individual preferences, product recommendations based on browsing or purchase history, and dynamic call scripts that adapt to customer needs in real time.

While AI provides the foundation for personalization, humans add the creative touch, crafting compelling stories that resonate with customers, adjusting messaging to align with cultural or situational nuances, and finding unique ways to engage customers beyond the data-driven suggestions.

CASE STUDY
Personalization in Action

A luxury car dealership used AI to analyze customer preferences and identify leads interested in electric vehicles. The sales team then invited these prospects to exclusive test drive events, blending data-driven targeting with a personal, creative touch. The campaign resulted in a 25% increase in test drive bookings and a 20% increase in sales.

BUILDING TRUST AND RELATIONSHIPS

AI contributes to trust by ensuring timely and

accurate communication, delivering consistent follow-ups that show attentiveness, and providing data-
driven insights that reinforce credibility. From there,
sales professionals deepen trust by understanding
and addressing customer concerns with care, being
genuine and transparent in interactions, and maintaining relationships even after the sale is closed. For
example, an AI agent might handle initial follow-ups
and provide basic information before a human sales
representative steps in to address more complex
questions, building rapport and trust.

ENCOURAGING INNOVATION
THROUGH COLLABORATION

AI can test different approaches, such as varying
email subject lines or call-to-action phrases, adjusting the timing of outreach campaigns, and
experimenting with different pricing strategies. Sales
professionals can interpret AI-driven test results and
apply their creativity to refine strategies further. For
example, they can use customer feedback to improve
messaging. Or, they can combine data insights with
personal anecdotes to enhance storytelling.

CASE STUDY
Innovation in Sales Campaigns

A fintech company used AI to test multiple lead nurturing strategies. Based on the results, the sales team

developed a hybrid approach that combined auto-mated educational emails with personalized phone calls, leading to a 40% increase in customer reten-tion.

Train your sales team to view AI as an ally, not a competitor. Provide hands-on training on AI tools to ensure they understand how to use AI insights to prioritize leads, how to collaborate with AI agents for seamless customer interactions, and how to inter-pret data and turn it into actionable strategies. Encourage a growth mindset by empowering your team to experiment with new AI-driven approaches, stay updated on the latest AI advancements, and em-brace continuous learning to remain competitive.

MEASURING SUCCESS IN THE HUMAN-AI PARTNERSHIP

When it comes to measuring success in the hu-man-AI partnership, be sure to evaluate the effectiveness of your human-AI collaboration by tracking lead conversion rates (measuring how well AI insights and human interactions work together to drive conversions and customer satisfaction scores). Don't neglect taking the time to monitor how quickly AI agents address customer inquiries. And fi-nally, be sure to calculate the ROI of integrating AI into your sales process.

Regularly review performance data to identify areas for improvement and ensure that both AI tools and human teams are performing at their best.

The future of sales isn't about choosing between humans and AI. It's about combining their strengths to create something greater than the sum of its parts. AI provides the precision, scalability, and efficiency needed to compete in today's fast-paced market, while humans bring the personal aspect customers value most.

By fostering a seamless partnership between human expertise and AI-driven capabilities, businesses can deliver exceptional customer experiences, build lasting relationships, and achieve unparalleled success. The perfect blend of human creativity and AI precision is not just the future of sales—it's the future of business.

In the next chapter, we'll take the final step toward mastering the AI-driven sales landscape by exploring how to build a winning AI-driven sales strategy that aligns with your business goals, drives revenue, and sets the foundation for long-term success.

CHAPTER 25

Building a Winning Sales Strategy for the AI Era

Artificial intelligence is no longer just a tool in the sales process—it's the backbone of modern sales strategies. Businesses that embrace AI can create sales operations that are smarter, faster, and more scalable than ever before. But a winning sales strategy in the AI era requires more than just adopting the right tools. It demands alignment with business goals, a focus on data hygiene and compliance, and a commitment to empowering your team to leverage AI effectively.

This chapter provides a step-by-step guide to building a successful AI-driven sales strategy that blends automation, personalization, and human expertise. By following these principles, you'll position your business for long-term success in the ever-evolving world of sales.

ADOPT THE RIGHT TOOLS

Before selecting AI tools, assess your sales process and identify areas for improvement. Important questions to ask include:

- Where are inefficiencies slowing us down?
- What tasks can be automated to free up time for sales reps?
- How can we improve personalization and customer engagement?

Integration is key. Make sure that your AI tools integrate seamlessly with your existing CRM and communication systems to avoid disruptions and maximize efficiency.

Implement a data hygiene process. Use a platform like Spearphish.io to audit your CRM regularly for outdated or incorrect records, append missing information, and verify customer preferences to ensure compliant and relevant outreach.

Next, be sure to train your team and empower your sales professionals. AI enhances human capabilities, but it doesn't replace them. Train your team to use AI tools to streamline workflows, interpret AI-generated insights to craft personalized strategies, and focus on building relationships and solving complex customer challenges.

Finally, commit to providing ongoing education. The AI landscape evolves rapidly. Keep your team up to date by hosting workshops and webinars on new features and best practices, encouraging participation in industry conferences and events, and creating a culture of continuous learning to stay ahead of the curve.

PERSONALIZE AT SCALE

AI enables businesses to deliver tailored experiences to every customer, no matter the scale. AI tools can analyze customer behavior to recommend products or services, personalized email campaigns can adapt messaging based on customer preferences, and dynamic chatbots can provide instant, customized responses to inquiries.

A sales rep might use AI insights to craft a pitch that resonates emotionally with a customer. Or, AI can suggest the right timing for a follow-up call. But it's the rep's empathy and storytelling that closes the deal.

Be sure you're engaging in continuous optimization, using performance data to refine your approach. Adjust your AI workflows based on lead conversion trends. Update scripts and messaging to address recurring customer objections. And experiment with new strategies, such as testing different outreach sequences or offer types.

Also be sure to use AI to drive strategic initiatives. Align your AI tools with broader business objectives, such as expanding into new markets by using predictive analytics to identify opportunities and enhancing customer lifetime value through personalized upselling and cross-selling.

With a clear plan, the right tools, and a commitment to continuous improvement, your business can thrive in the AI era and set a new standard for sales excellence. The opportunities are limitless. Embrace them, and lead your team into the future of sales.

Sales, as we once knew it, is gone. But what stands in its place is not an end—it's a rebirth. With the rise of artificial intelligence, sales is no longer about cold calls, cookie-cutter pitches, or relying on luck. It's about creating meaningful connections, anticipating customer needs, and delivering value at the right time, every time.

This book has explored the evolution of sales, the skills required to thrive in an AI-driven world, and the tools that can empower sales teams to achieve unprecedented success. From mastering data hygiene with Spearphish.io to leveraging the automation power of IgniteUps.AI, the path forward is clear: Embrace AI not as a replacement for humans, but as a partner that amplifies their strengths.

As we wrap up, let's revisit the core principles that will define the future of sales:

DATA IS THE FOUNDATION

AI is only as good as the data it processes. Clean, accurate, and actionable data is the backbone of AI-driven sales success. Platforms like Spearphish.io ensure that your CRM is optimized, compliant, and ready to fuel your AI tools.

"The quality of data is crucial for the success of any AI-powered sales strategy."

—*Elen Udovichenko, AI and Sales Technology Expert*

AI AND HUMANS
A WINNING PARTNERSHIP

AI excels at efficiency, precision, and scalability, but it cannot replace human creativity, empathy, and trust-building. The future of sales belongs to those who harness AI for repetitive tasks while focusing their human efforts on creating deeper, more meaningful customer relationships.

Customers demand more than a transactional

experience—they want to feel seen and understood. AI enables businesses to deliver hyper-personalized interactions at scale, but the human touch ensures those interactions are genuine and impactful.

CONTINUOUS LEARNING AND ADAPTATION

The AI era is dynamic, and staying ahead requires a commitment to learning, innovating, and adapting. Sales teams that embrace change, experiment with new strategies, and invest in their development will remain competitive in a rapidly evolving marketplace.

For decades, the best salespeople thrived on intuition, charisma, and hustle. Those qualities remain important, but the game has changed. The tools available today—AI-powered platforms, predictive analytics, and automation—allow anyone to achieve what once seemed impossible. The amateur can look like a guru. The small business can scale like a giant. The potential is limitless. But this opportunity is only available to those who act. Adopting AI isn't just about staying competitive—it's about leading the charge into a new era of sales.

The AI revolution is here, and it's transforming industries faster than anyone anticipated. You now have the roadmap to embrace this change and build a sales operation that thrives in this new landscape. Whether you're an individual salesperson, a team leader, or a business owner, the time to act is now.

Start by assessing your data, ensuring it's clean, complete, and compliant. Then, adopt the right tools. Platforms like Spearphish.io and IgniteUps.AI can provide the foundation for your AI-driven strategy. Take the time to properly train your team. Empower your people to use AI effectively while honing their human skills. Finally, continuously optimize. Measure results, refine strategies, and stay agile as technology evolves.

The future belongs to those who blend the precision of AI with the creativity of human expertise. It belongs to the businesses that prioritize personalization, compliance, and innovation. It belongs to the salespeople who never stop learning, adapting, and building trust.

This is your moment to lead the way into the next era of sales. Embrace the tools, master the strategies, and create a future where success is not just possible—it's inevitable.

The future of sales is here. Make it yours.

ACKNOWLEDGEMENTS

This book is the result of countless hours of research, collaboration, and inspiration, and it would not have been possible without the support and contributions of so many incredible individuals. While the words on these pages are mine, the ideas, lessons, and insights come from the collective wisdom and experience of those who've shared their knowledge and guidance with me.

To every salesperson who wakes up each day ready to build relationships, solve problems, and create value—you are the backbone of every industry. Your resilience, creativity, and dedication inspire the strategies and ideas in this book. This work is dedicated to helping you navigate and thrive in the ever-changing landscape of sales.

I owe a debt of gratitude to the mentors and educators who shaped my understanding of what it means to sell, to connect, and to lead. Your lessons on listening, persistence, and putting the customer first have been foundational to my journey. Thank you for showing me the art and science of sales.

To the many talented individuals I've had the privilege of working with, thank you for your collaboration and commitment. You've taught me the importance of teamwork, adaptability, and the power of collective effort. Whether it was in brainstorming sessions or the front lines of sales, your contributions have shaped my perspective and this book.

To the innovators behind AI tools like IgniteUps.AI and Spearphish.io, your groundbreaking work has set a new standard for what is possible in sales. The platforms you've created aren't just tools—they are transformative solutions that empower businesses to achieve extraordinary results. Thank you for leading the charge into the future of sales.

To those holding this book, thank you for trusting me to guide you through this journey. Your desire to grow, innovate, and adapt to the AI-driven era of sales is what inspires me most. I hope this book serves as a valuable resource and roadmap as you embrace the opportunities ahead.

To my family, whose unwavering support and encouragement have been the foundation of all my endeavors, and to my friends, who have been sounding boards for my ideas, thank you for always believing in me. Your love and patience make everything possible.

To the visionaries and pioneers who have redefined what it means to sell, your work has inspired this book and countless sales professionals around the world. The intersection of human creativity and AI precision is changing lives, and I am honored to explore that frontier alongside you.

Finally, I want to express my gratitude for the opportunity to write this book and share these ideas. Sales is not just a profession—it's a passion, a challenge, and an ever-evolving craft. As we step into this AI-driven future, may we continue to innovate, connect, and create value in ways that were once unimaginable.

Thank you for being part of this journey. Let's build the future of sales together.

With gratitude,
Chris J. Martinez

APPENDICES

The following appendices provide actionable tools, resources, and best practices to help you get started with AI-powered platforms like IgniteUps.AI and Spearphish.io. By leveraging these insights and tools, you'll be well-equipped to build a scalable, efficient, and personalized sales operation that thrives in the AI-driven era.

IgniteUps.AI is a revolutionary AI platform that automates inbound and outbound sales processes, empowering businesses to scale efficiently while maintaining personalization and compliance. Use this guide to quickly set up and optimize your IgniteUps.AI system.

Setting Up IgniteUps.AI

1. Sign Up and Onboard
- Create an account on the IgniteUps.AI platform.
- Integrate the platform with your existing CRM (e.g., Salesforce, HubSpot).

2. Configure AI Agents
- Define the objectives for your AI agents (e.g., lead qualification, appointment setting, follow-ups).
- Upload scripts, FAQs, and workflows tailored to your sales process.

3. Clean Your Data

- Before activating your agents, use a data hygiene tool such as Spearphish.io to clean, append, and verify your CRM data.

4. Launch Pilot Campaigns
- Test AI agents on a smaller segment of leads or customers.
- Gather feedback and refine workflows based on initial results.

BEST PRACTICES FOR USING IGNITEUPS.AI

Start Simple: Focus on automating repetitive tasks first, such as follow-ups and lead scoring.

Monitor and Optimize: Regularly review performance metrics and update workflows as needed.

Collaborate with Your Team: Train your sales team to collaborate with AI agents, ensuring a seamless customer experience.

KEY FEATURES TO LEVERAGE

Multi-Channel Engagement: Use AI agents to manage communication across email, phone, and SMS.

Compliance Tools: Ensure all campaigns adhere to TCPA regulations by managing opt-ins, DNC lists, and consent records.

24/7 Availability: Deploy AI agents to handle inquiries outside business hours, ensuring no lead is left unattended.

APPENDIX B
QUICK START GUIDE
TO SPEARPHISH.IO

Spearphish.io is a data hygiene platform designed to clean, append, and verify CRM data, ensuring that AI tools operate at peak performance. This guide will help you get started with Spearphish.io to optimize your sales data.

Setting Up Spearphish.io
1. Integrate with Your CRM
   - Connect Spearphish.io to your CRM (e.g., Salesforce, HubSpot, Zoho).
   - Allow the platform to access and analyze your customer records.

2. Run an Initial Audit
   - Identify duplicates, missing information, and outdated records in your CRM.
   - Generate a report summarizing the current state of your data.

3. Clean and Append Data
   - Use Spearphish.io to remove duplicates and errors.
   - Append missing contact details, such as phone numbers, email addresses, and physical addresses.

4. Verify Data Accuracy
- Confirm that all records are accurate and up to date, reducing the risk of failed outreach or compliance violations.

BEST PRACTICES FOR USING SPEARPHISH.IO

Schedule Regular Audits: Conduct monthly or quarterly data hygiene checks to maintain accuracy.

Focus on High-Priority Leads: Use clean data to prioritize leads that are most likely to convert.

Ensure Compliance: Use Spearphish.io's TCPA compliance tools to flag phone numbers on DNC lists and manage opt-outs.

APPENDIX C
AI IN SALES: COMMON
MISCONCEPTIONS AND FAQS

Misconception: AI will replace sales professionals

Reality: AI enhances, not replaces, human sales efforts. By automating routine tasks, AI allows sales professionals to focus on building relationships, solving complex problems, and closing deals.

Misconception: AI is too expensive for small businesses

Reality: AI Platforms such as IgniteUps.AI and Spearphish.io are cost-effective, scalable solutions that work for businesses of all sizes. Small businesses can benefit from AI's ability to handle high volumes of leads without adding headcount.

FAQ: How do AI tools handle compliance?

Answer: Most AI platforms, including IgniteUps.AI and Spearphish.io, include built-in compliance features such as:
- Managing DNC lists
- Automating consent tracking
- Adhering to SOC 2 standards for data security.

FAQ: Can AI tools integrate with my existing systems?

Answer: Yes, most AI tools are designed to integrate seamlessly with popular CRMs, email marketing platforms, and communication tools. Check with your provider for specific integration capabilities.

APPENDIX D
SAMPLE METRICS DASHBOARD FOR AI-DRIVEN SALES

Use this sample dashboard to track the performance of your AI-driven sales strategy.

Lead Response Time
Time taken to respond to new leads (ex. < 1 Hour)

Lead Conversion Rat
Percentage of leads that convert to customers (ex. 30%+)

Customer Satisfaction (CSAT)
Average satisfaction score from customer feedback (ex. 90%+)

Cost Per Lead (CPL)
Total cost divided by the number of leads generated $X (varies by industry)

Compliance Rate
Percentage of campaigns adhering to regulations (100%)

APPENDIX E
RECOMMENDED READING
AND RESOURCES

Books
- *Predictable Revenue* by Aaron Ross and Marylou Tyler
- *How to Win Friends and Influence People* by Dale Carnegie
- *AI Superpowers* by Kai-Fu Lee

Websites
- Spearphish.io: www.spearphish.io
- IgniteUps.AI: www.igniteups.ai
- HubSpot Blog: www.blog.hubspot.com

Courses
- "AI for Everyone" by Andrew Ng (Coursera)
- "Sales Strategy and Operations" by LinkedIn Learning

APPENDIX F
SAMPLE AI SALES WORKFLOW

OBJECTIVE: Automate lead qualification and follow-up for inbound inquiries.

Workflow Steps

Step 1

Capture Lead Information

- Use a form on your website to collect lead details (name, email, phone, and inquiry).

Step 2

Clean and Verify Data

- Use Spearphish.io to clean and append data, ensuring accuracy.

Step 3

Qualify Leads

- IgniteUps.AI analyzes data to score leads based on likelihood to convert.

Step 4

Automate Follow-Up

- IgniteUps.AI sends a personalized email within 1 hour of the inquiry.

Step 5

Escalate High-Value Leads

- Leads with a high score are routed to a human sales rep for personalized engagement.

Step 6

Track Performance

- Monitor metrics like response time, conversion rates, and lead engagement.

ABOUT THE AUTHOR

Chris Martinez began his journey in the automotive sales industry in 2003 with no clear roadmap. Despite having strong mentors, his growth came through hard-earned experience—falling, learning, and rising again in a highly competitive environment.

Since then, Chris has authored six bestselling books on automotive sales and marketing as well as one children's book. He is the founder of TheAutominer.com, a clean data marketing platform built for the automotive industry, which was acquired by PureCars in 2023. His entrepreneurial drive has also led him into ventures beyond automotive, including Spearphish.io and IgniteUps.ai.

Chris is passionate about helping ambitious salespeople and leaders think differently about how they engage with both current and prospective customers so they can consistently reach and exceed their marketing and sales goals.

ChrisJosephMartinez.com